T0178779

UNDERGROUND LEVIATHAN

MINING AND SOCIETY SERIES

Eric Nystrom, University of Nevada, Reno, *Series Editor*

Our world is a mined world, as the bumper-sticker phrase "If it isn't grown, it has to be mined" reminds us. Attempting to understand the material basis of our modern culture requires an understanding of those materials in their raw state and the human effort needed to wrest them from the earth and transform them into goods. Mining thus stands at the center of important historical and contemporary questions about labor, environment, race, culture, and technology, which makes it a fruitful perspective from which to pursue meaningful inquiry at scales from local to global.

Books published in the series examine the effects of mining on society in the broadest sense. The series covers all forms of mining in all places and times, building from existing press strengths in mining in the American West to encompass comparative, transnational, and international topics. By not limiting its geographic scope to a single region or product, the series helps scholars forge connections between mining practices and individual sites, moving toward broader analyses of the global mining industry in its full historical and global context.

Seeing Underground: Maps, Models, and Mining Engineering in America
Eric C. Nystrom

Historical Archaeology in the Cortez Mining District: Under the Nevada Giant
Erich Obermayr and Robert W. McQueen

*Mining the Borderlands: Industry, Capital, and the Emergence
of Engineers in the Southwest Territories, 1855–1910*
Sarah E. M. Grossman

The City That Ate Itself: Butte, Montana and Its Expanding Berkeley Pit
Brian James Leech

One Shot for Gold: Developing a Modern Mine in Northern California
Eleanor Herz Swent

Underground Leviathan: Corporate Sovereignty and Mining in the Americas
Israel G. Solares

Underground Leviathan

Corporate Sovereignty and Mining in the Americas

ISRAEL G. SOLARES

UNIVERSITY OF NEVADA PRESS | *Reno & Las Vegas*

University of Nevada Press | Reno, Nevada 89557 USA
www.unpress.nevada.edu
Copyright © 2024 by University of Nevada Press
All rights reserved
Manufactured in the United States of America

First Printing

Cover design by Louise OFarrell
Cover photograph: Smelter Smoke, Garfield. Used by permission, Utah Historical Society

Library of Congress Cataloging-in-Publication Data on file
ISBN 978-1-64779-136-0 (paper)
ISBN 978-1-64779-137-7 (ebook)
LCCN 2023942817

The paper used in this book meets the requirements of American National Standard for
Information Sciences—Permanence of Paper for Printed Library Materials, ANSI/NISO
Z39.48–1992 (R2002).

To Kristen

Contents

Acknowledgments

A book, a first book, is always the product of a network of trust and generosity that collectively pushes a project to completion. I am deeply grateful to the dedicated workers of: Archivo Histórico de la Compañía de Real del Monte y Pachuca, Archivo Histórico del Agua, Archivo Histórico del Estado de Hidalgo, Archivo Plutarco Elías Calles y Fernando Torreblanca, the Utah State Historical Society, the Salt Lake County, the Salt Lake Recorder, the Special Collections of the Harold B. Lee Library, the J. Willard Marriot Library, the Elmer E. Rasmuson Library, the Teresa Lozano Long Institute of Latin American Studies, the Center for the History of Medicine, the Library of the Harvard School of Law, the MIT Distinctive Collections, the Schlesinger Library, the Library of Indiana University Northwest, and the University of Colorado Boulder. This research was supported by funding and affiliation from Consejo Nacional de Ciencia y Tecnología, Secretaría de Relaciones Exteriores, El Colegio de México, University of Utah, Brigham Young University, Fulbright-García Robles Commission, University of Notre Dame, University of California San Diego, UC Mexus Institute, and the Weatherhead Center for International Affairs (Harvard University).

Different colleagues, mentors, and friends read different versions and pieces of this book in a multiplicity of seminars and institutions in different countries. I thank particularly the comments of John Womack Jr., Clara Lida, Mira Wilkins, Charles Maier, Sophus Reinert, Jaime Pensado, Graciela Márquez, Dan Graff, Elisabeth Köll, Darren Dochuk, Karen Graubart, Geoffrey Jones, Eric Van Young, Mario Barbosa, María Dolores Lorenzo, and Francisco Zapata. I am obliged to Jessica Barnard for the support in integrating into the Weatherhead Initiative for Global History and to Denise Wright for making the Kellogg Institute of International Affairs a second home in the last few years. I thank the friends that were vital in making me feel at home in the different transits. I give special thanks to Silvia, César, Quetzalcoatl, Maggie, Mandy, Jake, Melissa, Erin, Eden, Brenda, Ornela, Tim, Drew, Helen, and Abraham.

I am incredibly grateful to Susie Porter, Sven Beckert, and Mark Hendrickson for welcoming me into their academic networks, mentoring me in different stages of the research in the last eight years, and supporting me in over five dozen applications. Meeting my advisor, Aurora Gómez Galvarriato Freer, changed my life. Aurora always went beyond to support every step of the process, and her comments shaped the main ideas of the first version of this research. I will be forever in debt to Ted Beatty. He read the dissertation in Spanish and the rewritten version carefully multiple times and always had kind suggestions, from the general argument to too many grammar and word choice issues. Ted's conviction and work for this project to be published pushed me to move forward in the difficult years after I completed my PhD. I will forever cherish the kindness of all of them in understanding the challenges, both personally and academically, of a historian with a PhD from a Mexican institution trying to write a continental narrative from the South. I thank Eric Nystrom for his trust in the Mining and Society series project and the University of Nevada Press editorial team for the absolute wonder of working with them during the last few years.

During the multiple travels that this book required, I met my life partner. Kristen's love has supported me through the hurdles and obstacles that followed once I finished my PhD. Thank you for everything.

UNDERGROUND LEVIATHAN

Introduction

> Probably no mining company of the world has had more to do with the development of the metal industry than the United States Smelting, Refining & Mining Co.
>
> —*Romances of Great Mines, number 39*

This is a book about the people inside and around a mining corporation that moved and ruled the underground world: the United States Smelting, Refining, and Mining Company, widely known as United States Company. There was nothing extraordinary about the firm, but, precisely because of that, its history is relevant. It formed part of a global phenomenon, the emergence of mining behemoths in the Americas in the late nineteenth to early twentieth century, which ruled the fate of thousands of individuals, communities, and states. As one of the five biggest nonferrous mining corporations at its time, the US Company is a window into an underground world that transformed industrial societies on the surface.

Ordinary does not mean small or insignificant. The US Company coordinated the location, extraction, smelting, transportation, refining, and sale of minerals across the Western Hemisphere, from Alaska to Peru, and controlled the fate of thousands of workers, managers, engineers, peasants, politicians, and neighbors in dozens of communities. The firm was engaged in all the branches of the mining business, from the search of minerals to the sale of consumer products. "Its business," argued the *Mining Outlook,* "is so diversified and so extensive that the company is a factor in the production of every individual metal of the country, except nickel, and has its properties located in almost every mining camp of the world."[1] At the peak of its relevance, the United States Company produced 7 percent of the gold, 7 percent of the lead, and 13 percent of the zinc extracted in the United States, besides mining over 50 percent of the gold and 30 percent of the silver in Mexico, equivalent to 13 percent of the global production of the white metal.[2] The metals produced by the firm were used as currency and weapons, as paint and pesticides, as jewelry and as fuel, in photography and in electric circuits. The powerful organization ranked, along with six other metallic mining companies, among the twenty most valuable firms in the United States in 1917, exceeding the value of the companies in petroleum, gas, coal (and their products), transportation equipment, food, and rubber combined.[3]

The main argument of this book is that multinational corporations are not solely economic organizations and that their political functions are not merely accidental. The multinational mining firms during the first decades of the twentieth century were a combination of human labor and ingenuity that transformed matter in the earth into both valuable minerals and toxic waste. The actors that intervened in the corporation established a multinational model of business in the Americas, long before the emergence of the multidivisional distribution and retail firms. The complexity of their operations empowered a class of engineers and technical experts that made the main decisions in mining communities across the continent and who rapidly innovated new forms and practices of business operations. They were at the forefront of technological innovations in that era. While meat companies in Chicago killed cows by hand, American multinationals in Chile operated high-technology smelters and established research and development units; and long before Ford installed the assembly line on Highland Park, silver cyanidation plants in Mexico used carefully calibrated concentration and refining chains and flows.

I began my research looking for the influence of a prominent capitalist that could lead the narrative of the firm's history, but he could not find a single person with those characteristics. Instead, he found hundreds of minor actors, with medium responsibilities, which wielded the visible hand, or the iron fist, in the organization of the corporation. The high technological development of mining likewise implied that mining engineers, geologists, and chemists were the main actors in the construction of the companies. Some investors had a role in the formation of the corporations, namely, the Guggenheim and the Rockefeller families, but this book proposes that middle and top engineers were responsible for the design of the technological and organizational features of these multinational titans.

To be sure, valuable products were not the only objects that resulted from the operations of the firms. The corporation transformed rock into the infrastructure of modernity, but also into diseases in the lungs of the workers, and into poison in the fields around its plants. Moreover, the corporation was more than just an instrument of shareholders or of management. The transformation of the underground and surface world caused by the United States Company involved not only capitalists and consumers, but also a wide variety of technicians, workers, scientists, organizers, politicians, farmers, and neighboring communities. In short, this book argues that the corporation cannot be reduced to an economic organization, producing merchandise and generating profit, nor did its actors simply act to maximize the value of company shares. Indeed, the corporation incorporated individuals, power relations, and matter itself: it was a Leviathan.

On April 10 and 11, 2018, the world was reminded of the modern influence of corporate behemoths, as Mark Zuckerberg testified in front of the Congress of the United States. The hearing took place a month after *The Guardian* revealed that a political communication firm, Cambridge Analytica, had access to the profiles of millions of Facebook users and had weaponized the data to create misinformation during the presidential elections of 2016. The hearing exposed the generalized technical ineptitude of the representatives and their inability to design any regulation of the social media giant. Moreover, it revealed an ecosystem of corporate entities that could compete with the state. "In a lot of ways, Facebook is more like a government than a traditional company," declared Zuckerberg to journalist David Kirkpatrick. "We have this large community of people and, more than other technology companies, we are really setting policies."[4] Facebook is not alone in this quest for governing; for example, Amazon collects taxes and coordinates international surveillance, and Blue Origin, a company also owned by Jeff Bezos and used by Amazon to send their satellites, plans to build extraterrestrial colonies.[5]

The sovereign of tomorrow appears to have "Corp." as its last name, but the power of corporations has a long history. The structure of the western European expansion into the world in the seventeenth century relied heavily on the existence of corporations rather than on the direct state power of the crown. In the late eighteenth century, Edmund Burke denounced the power of the East India Company. He argued that only the exterior body of the firm was business-oriented, as the inner workings of the corporation had all the features of a sovereign entity: it was a "state disguised as a merchant." The firm had claims over territory, rules, magistrates, and subjects. Formally, the firm's power derived from two sources, the charters with the crown of Great Britain and the various charters from the Mughal Empire. Nonetheless, the firm's sovereignty was the only legitimate union between the two empires, as the corporation was not a part of the British nation but an independent deputation of individuals. These covenants entered the firm's service and followed its orders, but it was a body entirely created by authorities: "a commonwealth without a people...a kingdom of magistrates."[6] The commercial origins of the firm hid the true sovereign nature of the organization, and neither the state nor the people were able to keep the power of the corporation in check. To use Burke's words, "The constitution of the Company began in commerce and ended in empire."[7]

The same debates were repeated two centuries later, during the emerging dominance of US firms in the Americas. The term "Banana Republic," created by O. Henry in 1904, described the immense power of US corporations and their managers in countries dominated by agricultural exports in the Americas. In one of the short stories of *Cabbages and Kings,* the minister of finance of the Republic of Anchuria, Mr. Espirition, tries to negotiate taxes with the representative of the Vesuvius Fruit Company, Mr. Franzoni. After an argument, Espirition declares that the offer of the company insults the government. "Then," said Mr. Franzoni, in a warning tone, "we will change it." While the offer is never changed, the company decides to remake the administration of the country. The story was not far from reality, as corporations were the real agents of American colonial power across the continent. The United Fruit Company that inspired the book by O. Henry lobbied for the Guatemalan coup d'état in 1954, and the American firms operating in Chile pushed for the Coup of Pinochet in 1973.[8] The journalist Jack Anderson described one of the firms behind the Chilean coup, the International Telephone and Telegraph (ITT) Corporation, as a "veritable nation, a gigantic conglomeration of interests throughout the world. The ITT is like the British Empire was: the Sun never sets in its domains."[9] The comparison with the British Empire by Anderson foreshadowed the continuity between colonial powers and the operation of modern firms.

This book argues that the power wielded by these corporations was not extraordinary but a structural phenomenon in capitalism between 1870 and 1970. It examines the emergence, dynamics, and consequences of one of these corporations during the first half of the twentieth century. The United States Company was incorporated in Maine in 1906 and held properties in Utah, Colorado, California, Nevada, Alaska, Mexico, and Canada—and tried to expand to Honduras, Venezuela, and Peru. The firm carefully planned its activities across productive, technological, economic, urban, environmental, political, and cultural spaces. It shaped, and was shaped by, international actors: managers, engineers, workers, neighbors, and farmers. The firm, as a modern behemoth, was not only a technological machine, or an economic organization, but also a political body that had objectives beyond the interests of its shareholders. If the Dutch, British, and French East India Companies were vehicles of merchant capitalism in the early modern period, the US multinational firm was the primary engine of the machinery of the United States' corporate empire, first in the continent and then in the rest of the world.[10] This books tells the history of an exemplary US multinational firm and in doing so opens a new horizon to understand the dynamics, actors, and nature of corporate power in the twentieth century.

Why is the history of corporations not part of mainstream political history? General histories rarely engage the emergence of these machines of global integration. National-territorial states remain the focus of most political histories, which neglects the existence of global, non-territorial, and non-state sovereignties. In the cases where the political relations of the firm appear, they are often discarded as appendices to the merchant drive and profit motive of the firm, or a product of the corrupt and collusive nature of the state. In addition, the study of corporate histories is more typically reserved for those initiated in business history, and, more often than not, it is restricted to the economic and organizational dimensions of the corporate organism.

In the last decade, new research on the political characteristics of firms has explored issues previously ignored by traditional national and business histories. In particular, two sets of studies on exemplar companies have examined the many layers of social interactions coordinated by the corporation. In the first place, the monographs centered on the operation of the East India Companies and their role in the definition of colonial power. Over a century ago Frederic W. Maitland established the relationship between the sovereignty of the British East India Company and the sovereign claims on territory on the Americas.[11] New legal evaluations of chartering mechanisms and corporate laws examine the transformation of the medieval institution of the corporation as a continuous form of legal sovereignty on private spaces throughout many centuries.[12]

New histories of the East India Companies have revived Maitland's argument, demonstrating the multiple sovereignties competing in the early modern period. Trading companies used the medieval institution of the corporation and competed with nation-states and traditional kingdoms in the definition of political bodies around the globe. These studies show that the empires of Western Europe and the global market did not emerge exclusively as a project of the crowns of Western Europe, but also from the collegiate institutions of the trading firms.[13] Furthermore, this realm of diffuse political power did not end with the establishment of strong European states in the late eighteenth century, as new private sovereignties emerged once again in the late nineteenth century in Africa.[14]

The second group of studies on corporate sovereignty revolve around the influence of the United Fruit Company.[15] The histories of this corporation mirror the main arc of the early modern trading firms. As with the East India Companies, the operation of the United Fruit Company shaped the form of different states under a single empire, although the banana organization did not face substantial competition from other corporations. Started primarily as a trading firm, the company increased its control over territories over time and consequently faced growing levels of resistance in the Caribbean, Central America, and the United States. Eventually, like the British East India Company, the United Fruit Co. restricted its operations to the merchant businesses, leaving the overt territorial intervention in the hands of the United States war machine.

Recent literature on the political history of corporations has greatly improved our knowledge of the dynamics of power of the firm with a variety of actors: local elites, imperial states, worker's organizations, and cultural apparatuses. Nonetheless, studies of American corporations of the twentieth century have three major shortcomings. First, they have had only episodic access to the internal sources of the firm, and therefore the narrative of most stories relies heavily on some correspondence from the local subsidiaries and documents produced by national or local government agencies.[16] This is especially clear in the case of the United Fruit Company, as its histories are situated in different nation-states (Guatemala, Costa Rica, Honduras, Ecuador, and Colombia), with limited parallels and connections between national contexts. While these monographs give a clear picture of the national context, and the tensions between the headquarters and the republics where they operated, it is remarkable that there is little resemblance in the corporate outcomes in an international level. In other words, by structuring the analysis by nation-states, the multinational empire appears as a sum of host nations under a single home country, not as an integrated international corporate organization.

The limitation due to the sources leads to a second analytical shortcoming. For the most part, the corporation appears only as an instrument for the exercise of power by meta-forces: for instance, American imperialism, white supremacy, the global market, US consumer culture, and the shareholder's benefit.[17] To be sure, all these underlying causes helped define the firms' structure, dynamics, and strategies, but none of them can fully explain the firm's actions. As studies center their accounts wholly on these external and contextual factors, the corporation itself remains a black box, opaque in its structure and operations, and simply a vehicle for larger political or social actors. The classical dichotomy between host nation and home country in the history of multinational corporations hides the multiplicity of local, national, and transnational social agents, acting inside the internal structure of the firm as well as disputing their role in other political spaces. In contrast, this book argues that more than an actor in an outside political arena, the firm was a political result of social conflict and coordination.

By analyzing corporate archives of the United States Company across its major divisions, and using local sources from multiple actors, this book offers a fundamentally different perspective on the nature and dynamics of the emerging corporate multinationals of the period. That is, the unit of analysis of this book is not the home country, or the host nation, but the corporation as a single global entity. This approach shows that the power of the firm was not a mere accessory to the profit motive or to the imperialistic drive of US capitalism, but a social result constituted from many global, national, and local forces. The history of the incorporated body of the firm was not simply the history of its founders or its shareholders, nor was it the history of the political interests around it. The analysis of the United States Company in this book is not therefore constrained to the legal and formal attributions of the firm, but extends to the myriad interactions between workers, managers, farmers, and consultants during the critical years of the birth of the multinational firm in the Americas. The relations of these social actors, even while disputing the power of the firm, constituted the corporation as a sovereign entity.

To be sure, this book does not suggest that the state's power was irrelevant or that the sovereignty of the corporation always superseded the influence of empires or the nation-state. It is a truism to remark that the British Empire transcended the East and West Indian Companies, or that the nation-states in Latin America existed before the arrival of the big US multinational corporations, such as the United Fruit, and kept their essential territorial unity after their dissolution. Nevertheless, the structure of these states, like Guatemala and Chile, changed after the firm's influence, and, while individual corporations perished, the corporate system and its power did not. In other words, more than asking if sovereignty is a characteristic either of states or corporations, or who has the more stable or profound power over its subjects, this book is interested in the simultaneity of different sovereign powers with distinct characteristics. More anonymous and numerous corporations wielded sovereign powers in the Americas, co-creating the general structure of sovereignty—imperial, national, and corporate—in the continent.

To be sure, corporations have not been altogether absent in histories of political sovereignty. In his *Leviathan,* Hobbes analyzed the constitution of body politics among the merchants, inside a family, outside the limits of the commonwealth, and inside the commonwealth itself.[18] These bodies, he argued, were capable of personhood, because they could collectively defend their interests in the law and establish their own internal rules. Moreover, Hobbes identified the deep relationship between monopolies and corporations: the incorporation of a body of traders implied the formation of a single buyer of the products of the colonies and single seller of those merchandises in the metropolis. In other words, Hobbes's classical study did not erase the existence of other incorporated, artificial bodies inside the commonwealth, such as the British East India Company, but warned against the possible contradictions derived from their competing powers. He considered that the double monopoly of trading firms was detrimental to the English commonwealth, and argued that a proliferation of corporations with distinct sovereign powers was an infirmity: "as it were many lesser commonwealths in the bowels of a greater, like worms in the entrails of a natural man."[19]

Since the times of Hobbes, the metaphor of corporations as monstrous creatures has not typically taken the form a worm, but of a giant octopus: a kraken. The cephalopod has incarnated the corporation as it exemplifies a capacity to move its tentacles with relative autonomy while maintaining a general coherence. As the head commands and devours everything within reach of the body, it has exemplified the center-periphery and insatiable nature of multinational corporations.[20]

Although this portrayal has reappeared over and over in debates about the characteristics of the firm for centuries, it has some limitations compared with the original Hobbesian metaphor. The image of a kraken reinforces the idea of an organic being with a centralized volition. Surely the Leviathan, whether depicted in the Bible or in Melville's *Moby-Dick,* can also represent an organic beast capable of devouring men. Hobbes's Leviathan, however, is an artificial being, an automata, and therefore a designed creature. The body of the corporation, in consequence, is not the product of a predetermined organic growth, yet neither is it solely an artifact.[21] Its volition is not centralized in a single head, nor is it the vessel of an external prerogative. It has not swallowed the humans: it is, in fact, the humans.

The second advantage of the traditional Hobbesian metaphor over other ways to analyze the modern corporation is that, although the Leviathan is unique, the political body of the corporation represents a multiplicity of interests, agencies, and social actors. For Hobbes, the commonwealth is instituted by covenants and not defined a priori by a precedent power. This implies that the multitude of men, or their representatives, establish an alliance to grant the power of government to a man or an assembly of men. In that sense, the resulting sovereign is not a beast that has devoured the individuals, but it is an artificial being entirely composed by them. Its actions are the product of the interaction of its covenants' volition. At the same time, these covenants are the very same means of the exercise of sovereign power, as they are its subjects and magistrates. "Leviathan . . . is one person, of whose acts a great multitude have made themselves every one the author."[22] The unity of the structure for Hobbes only reflects the multiplicity of its composition; the Leviathan is not a single will but a social result. The corporation is also a product of multiple agencies of social actors interacting inside the political body.

The third advantage of using the Hobbesian definition of the Leviathan is that it allows for the expression of multiple bodies within one. By following the Hobbesian concept of the *fictitious bodies* of the sovereign, it is possible to describe different layers in the firm's operation without assuming a perfect harmony between them.[23] The corporate Leviathan contains multiple bodies and their representatives (or covenants) that collaborate in the division of labor inside the firm. Managers, technicians, providers, workers, neighbors, and political actors work together in the operation of the firm but also develop distinct organized bodies inside, like unions, professional organizations, and social networks. Their existence is neither parasitic of a primary political body, such as worms in the bowels of a greater man, nor do their actions act in harmony with all the other members mastered by a single volition, such as the tentacles of the kraken. In contrast, the fictitious bodies expand and contract, the result of design, redesign, and conflict inside the corporation. In the US Company, as in most modern corporations of its day, some of the bodies also developed a consciousness of the power of the sovereign and their place as members of the artificial being.

By revising the Hobbesian metaphor, this book presents the history of the formation and design of the fictitious bodies of the United States Smelting, Refining, and Mining Company. Each of the bodies analyzed uses one of the three primary analytical axes of the automatic organism used by Hobbes: *Matter, Form,* and *Power.* In this book, Matter explains the material characteristics of the US Company's operation. Mining corporations transported several materials from the ground to the markets as products but also as pollution in the environment. The Form of the Leviathan refers to the structure of the firm, in particular, the organizational features pushed by the technicians and managers. Finally, the Power of the corporation examines the political features of the firm and the dispute over the characteristics of the sovereign body between the main antagonists: management and labor.

The reader should forgive the book for not offering a traditional emergence-development-demise structure that would allow a straightforward narrative of the history of the corporation. During years of research it became clear that the realities of the firm were both synchronical in the multiple divisions and asynchronical in the different layers of operation. The firm had multiple temporalities within, which hid the totality of the operation for different actors on the inside. The analytical approach of this book is to connect the formation of different fictitious political bodies of the firm, from the top managers to unskilled laborers, to the material realities that they articulated, from the creation of wealth from rocks to the contamination of rivers with waste. These histories might seem foreign to each other, and, as matter of fact, they often appear as such in traditional local and national narratives. Nonetheless, this book foregrounds how the corporation interconnected the multiple realities on a layered, planned blueprint. The six chapters explore multiple paths that led to the Leviathan, asking about the organizational characteristics of the firm, the Form; its political structures, the Power; and the material realities of the mining business, the Matter. Each chapter will follow the emergence and dynamics of one of these bodies in order to draw a complete image of corporate sovereignty.

Chapter 1 addresses the question of the *Form* of the Leviathan by examining the duality of the artificial personhood: the corporation as an organized body. Beginning with the incorporation constitution, in Maine in 1906, the narrative describes the main moving parts of the firm. It argues that the ghostly character of the incorporated body allowed the territoriality of the firm in multiple places in the continents from Peru to Alaska, and the circulation systems established by the holding company offices in Boston.

Chapter 2 refers to the question of the *Power* of the Leviathan: the corporation as a body of subjects that wields control over individuals. The chapter examines the composition of an international body of workers within the firm and the actions of management to establish and extend its authority over them. The chapter argues that the ways in which industrialized mining increased corporate control over the work process determined the configuration of a new worker's organization, which replaced the traditional forms in the first decades of the twentieth century. The chapter analyzes the evolution of simultaneous labor strikes within several divisions of the firm, and how they transformed the organization of work inside the corporate Leviathan. Increasing labor unrest in the 1920s and the changing nature of the body of working subjects triggered the establishment of novel methods of control by corporate management. Technocrats within the firm engineered new relations between the corporation and its subjects, through the emergence of labor departments and the standardization of medical controls in the early 1930s.

The question of the *Matter* of the Leviathan is treated in chapter 3: the corporation as a mineral body. Mining companies transform substances found under the ground into useful refined components for industrial purposes. The chapter traces several phases in the detection, extraction, and processing of these materials through the work of different consultants hired by the corporation, from the sulfide ores in the mountains of Real del Monte to the detection of silicon sulfide particles in the lungs of miners in the company's hospital. The chapter argues that the location and use of Matter in the corporate body of the firm not only impacted the creation of wealth in the US Company but also was instrumental in the conformation of the university and consultancy clusters in the East Coast of the United States. In particular, the chapter focuses on the relationships established by the firm with consultants in Boston, both at Harvard and MIT. The final result of the incorporation of a body of technicians into the firm was a detailed account of the corporate's material body, the minerals, and its relations with the physical bodies of the workers.

Returning to the question of *Form* of the Leviathan, chapter 4 looks at the corporation as a body of sovereigns, or a "Republic of Magistrates" as described by Burke. Following the observation of Thorstein Veblen on the capacity of technicians of forming their state, "a soviet of technicians," the chapter describes the formation of a technocratic global bubble for upper management in the US Company. It argues that the establishment of a system of transfers of managerial personnel among divisions of the United States Company, and the publishing of a corporate magazine, created an American managerial community specific to the firm. The effects of these policies were not solely the creation of an internal corporate culture, but also the emergence of a unified experience of the American colony for the engineers of the US Company.

Chapter 5 reexamines the *Power* of the Leviathan, describing the formation of contractual bodies of workers in Mexico and the United States and a growing integration between their organizational strategies. It continues the story of the organized workers inside the firm, and their relationship with the mass unionization policies adopted in each country in the 1930s: of the New Deal and of Cardenismo. The organization of laborers by radical unions pushed the boundaries of national sovereignty across North America, reaching the distant sections in Alaska and forming alliances transnationally. These frictions with other sovereignties, corporate and state, resulted in a forced dissolution of the international body of workers by the House Un-American Activities Committee (HUAC) trials in Salt Lake City after the Second World War.

Chapter 6 evaluates the question of the *Matter* of the corporation, analyzing the firm as a body of waste. It engages with two different debates: the business history debate about the limits of the firm, represented by the economist Ronald Coase; and the natural state of the Hobbesian tradition.[24] This chapter argues that the definition of what was outside and inside of the firm was not only an economic decision, but also a political one regarding the sovereignty of the firm. The corporation established sovereignty outside its property rights, building a *space of exception,* an invisible body of the firm. It focuses on the conflicts of the company with Mormon farmers in the Salt Lake Valley and indigenous communities in the Huasca Valley. After decades of struggle over the effects of sulfides and industrial mining in both places, the firm developed farming experiments trying to prove the beneficial effects of sulfur dioxide, arsenic, and cyanide on its neighbors. The expulsion of these materials into the environment was an expression of an invisible body of the firm, and its indirect sovereignty over territory, that allowed the ex-corporation of matter and the effective expansion of its power.

The structure of this book follows the many layers and structures of the US Company, the diverse locations occupied by the firm, and the multiple forms of sovereignty exercised by the organization. Memory of the firm's activities appears in the records, land, and anecdotes in the various localities, but the consciousness of a single transnational history is always hidden. This perpetual exogeneity of the firm, in the US Company as in many corporations, concealed the territoriality of its operation and the sovereign nature of the organization. In many ways, the research is a collection of case studies linked by a single actor, an abstract and ghostly firm, which always appeared alien to local spaces. The chapters and their subsections reflect the fragmentation of these stories, but also the global nature of local actions taken by workers, peasants, engineers, scientists, and politicians. The actors of this book inhabited both a cosmopolitan and a nationalist world, like the one we live in today, but on occasions were aware of their global connections. Most of the time they could distinguish the face of a common sovereign that governed extraterritorially, but only episodically they gained consciousness of their concealed rule: they were the Underground Leviathan.

CHAPTER ONE

The Ghost and the Machine

From time to time, after a few answers had been given, the exhibitor would apply a key to the Turk's left side, and wind up some clockwork with a good deal of noise. Here, also, he would, if desired, open a sort of lid, so that inside the figure you could see a complicated mechanism consisting of a number of wheels; and although you might not think it probable that this had anything to do with the automaton's speech, it was still evident that it occupied so much space that no human being could possibly be concealed inside.

—E. T. A. Hoffman, Automata

PORTLAND, MAINE

In 1846, Henry David Thoreau followed the lumber trade and made excursions far north with some Bangoreans. The transcendentalist and his companions stopped by ferry at Indian Island, part of the Penobscot territory according to the treaty of 1819 between the tribe and the Commonwealth of Massachusetts. By then, the tribe had only 362 members. "This picture will do to put before the Indian's history, the history of his extinction," wrote Thoreau in his diary. Despite the destruction of the community, the territory seemed populated with ghosts. He observed a few new houses in the many miles of land, "as if the tribe had still a design upon life."[1]

The Penobscot were part of the Wabanaki (People of the Dawn) Confederacy. Out of several peoples populating the area of northeastern Massachusetts and Acadia, the Wabanaki had reimagined themselves after arrival of the white colonizers and could establish strategic alliances with English and French settlers. The Wabanaki imposed their power over the rivers, against the dominance of the Massachusetts colony, explored farther lands, and challenged French rule farther north. They fought to preserve their sovereignty against the Anglo settlers, fighting six wars for the frontiers of Maine between 1670 and 1760, frequently defeating, and even humiliating, the colonizer forces. Nevertheless, the tribes had a hard time containing the white settlers after the independence of the colonies. The white independent forces broke the confederacy into smaller units and incorporated the state of Maine in 1820. The Wabanaki tribes continued to occupy some territory and settled on the internal islands of the rivers. Ghosts of this shrinking sovereignty appeared as a newly domestic space in the forest's wilderness. Spirits populated the mountains, the rivers, and the forests during the excursions of Thoreau with Wabanaki guides.[2] When, in one of his last trips to the area, Thoreau asked his guide, Polis, if he was happy to return home in Oldtown, the proud Penobscot answered, "It makes no difference to me where I am."[3]

Excursions such as Thoreau's were a common exercise among the white population of northern New England as they tried to establish symbolic and material frontiers between their claimed territories and the British possessions in Canada. The maritime space was also changing for the white settlers in Maine. To compete with Boston and New York, the authorities of Portland—then Maine's commercial center and first capital—established trading ties with the cities of Quebec, integrating into their commercial sphere. These golden years for Portland ended with the Great Fire of 1866. Thereafter, Maine dealt with the ghost of peripheral existence once more. The city abandoned a project to transform into a railway hub and embraced its secondary place until the end of the century. In 1908, another fire destroyed even those phantoms of progress.[4]

I am looking for a different ghost: a mining corporation. On January 9, 1906, the United States Smelting, Refining, and Mining Company was organized in Portland under the laws of Maine. As I look for traces of it on the records, I find that it was organized as a democratic organ of its owners, the shareholders. As a Limited Liability society, owning shares of the firm did not represent a personal obligation for shareholders, and they had no role in the decisions of the firm. Different shareholders had different rights to the profits of the company, but they all depended on the authorization of the board to obtain their dividends. The general assembly was the moment of deliberation in this republic of owners, one-share-one-vote, but most of the corporate constituents relegated their voting rights to the managers. In short, the formal owners of the firm appeared as an abstract entity, an interlocutor that never answered back, an absent spirit in the spaces of the firm. Companies controlled by the personalities of a single tycoon were the exception, not the norm.[5]

Shareholders owned the firm, but, more relevantly, the corporation owned other firms. In 1916, the US Company had the majority or the totality of shares in fourteen companies. Each one constituted a legal person on its own, with different constitutions sorting its subjects, forms of government, and the capacity of voters. An individual firm could declare bankruptcy, decide on its dividends and profits, invest, trade and contract debt, and so on. Five were incorporated outside Maine: the US Stores Company and the San Pete Valley Coal Company in Utah; the US Metals Refining Company in New Jersey; the Compañía de Real del Monte y Pachuca in Mexico; and the Carbon Emery Stores Company in Nevada. Nonetheless, the majority of the firms of the US Company conglomerate were incorporated in Portland: the US Smelting Company, the Centennial Eureka Mining Company, the Mammoth Copper Mining Company, the Gold Road Mines Company, the Needles Mining and Smelting Company, the U.S.S.R.M. Exploration Company, the Richmond Eureka Mining Company, the Niagara Mining Company, and the Utah Company. This last subsidiary owned all the shares of five additional firms: the Utah Railway Company, the Castle Valley Coal Company, the Consolidated Fuel Company, the Black Hawk Coal Company, and the Panther Coal Company.

Yet none of these, neither owners nor managers or technicians, left any trace in Maine. Nowhere in the state's history, full of wars, indigenous tribes, pirates, and Irish Longshoremen, did mining companies have any role. While surveying the boundaries of the state, Joseph Treat found deposits of usable quarry up north. An extensive examination of the volcanic rocks on the eastern border with Canada found no signs of metals, only large amounts of fossils.[6] The exploration of Maine's glacial gravels offered the same results, unable to find any trace of usable metals deep into the old Wabanaki stronghold. By 1907, only one mining company, the Mount Glines Gold & Silver Mining Company, operated in the state, on the eastern border with Canada. Employing only twenty-five men, it could only hope to get some smaller deposits compared with the minerals in the neighboring territory of the Mi'kmaq.[7]

The only trace of the firm in Maine is in its incorporation record, which made the firm a Maine citizen in the eyes of the government. Twenty years before, the ruling of the US Supreme Court in *Santa Clara County v. Southern Pacific Railway Co.* defined corporations as legal persons: "The Court does not wish to hear argument on the question whether the provision in the Fourteenth Amendment to the Constitution which forbids a state to deny to any person within its jurisdiction the equal protection of the laws applies to these corporations. We are all of opinion that it does."[8] The decision created a special kind of citizen that could decide where and how to be born. Furthermore, the ruling of the Supreme Court produced the miracle nature could not: creating mining wealth in Maine, at least on the paper.[9]

In 1902, the manual of *Comparative Advantages of the Corporation Laws* considered that the corporations would "naturally make [their] choice between Arizona, Delaware, Maine, New Jersey, New York, South Dakota, and West Virginia, all of which have made a great effort through advertising and voluminous literature to attract the increasing corporation business of the country."[10] In 1912, at least seventy-three mining and smelting corporations and fifteen railroad firms were incorporated under the laws of Maine, operating from Alaska to Argentina. Maine offered low costs and not a single requirement of presence in the state. Stock and transfer books could be kept anywhere; no records were necessary, though lists of stockholders or statements had to be public; there were no limits to capitalization; firms could start a business with only three shares subscribed, with absolute freedom from personal liability; and, even if the assemblies had to take place inside the state, there was no restriction on using proxy votes.[11] In other words, the state that chartered the corporation had little to do with the real operation of the firm. The corporation and its managers were free to be ghosts in Maine, at a minimum cost.

Although Maine and New Jersey were especially liberal in their incorporation laws, firms around the world faced very few restrictions in their operations. In Mexico, for instance, firms had similar freedom according to the Commerce Code (Código de Comercio) of 1889. It required only five shareholders to form a Sociedad Anónima (Limited Liability Corporation), and they did not have to publish their balance sheets. Much like in Maine, the general assembly had to take place every year, and the board had to have sessions quarterly, but they did not have to keep the records in Mexico City and there was no limit to the use of proxies.[12] Just as in the United States, the proxies had to have at least one share in the company, and, in consequence, the managers became owners of one share each. This convergence in the incorporation laws allowed firms to maintain an international existence, independent of space, state sovereignty, and shareholders' ownership. Instead of location and physical presence, the unity of the corporation relied on a system that organized ownership, management, work, and technical control.[13]

The wide set of aims and goals of the corporation defined in the act of incorporation was the closest the behemoth could aspire to obtain a soul. The theory of the Leviathan of Hobbes was deeply influenced by the exhibition of automatons in Paris in the end of seventeenth century. Those creatures, pulled by invisible strings, influenced René Descartes's dualist theory of body and soul. British philosopher Gilbert Ryle later described the premise of Descartes' dualism as "a ghost in a machine."[14] In the analogy of the corporation, the ghost of the corporate citizen relied on the incorporation of the firm. The US Company existed only in this way, immaterial, in Maine, but the rest of the components of the firm were a material machine: from the cranes, crushers, and locomotives to the administrative procedures, evaluation mechanisms, and promotion schemes; from the furnaces, dredges, and sintering machines to the labor policies, environmental strategies, and political negotiations. The rest of this chapter will outline the key features of the productive organization of this automaton, following the emergence of the firm's operations in Utah, its expansion to Mexico, the operations in the East and Midwest, and the ever-expanding frontiers of the firm. The constituent parts of the US Company—each incorporated firms themselves—were distinct operations from each other, but they were pulled together by the strings of the holding. As the automatons in Paris in the seventeenth century, they were all designed on the working table of engineers.

BINGHAM CANYON, UTAH

In 1906 Albert Holden, the architect of the firm, was still in his thirties but was not a newcomer to the mining business. He had inherited a prosperous mining business in Bingham Canyon. Half a century before, when the Bingham brothers discovered metal in the Oquirrh

Mountains, Brigham Young forbade them to exploit it, fearful that the notice of mineral richness would attract unwanted gentiles into the promised land. Liberty Holden, Albert's father, was one of those unwelcomed miners in search of fortune. A professor in Cleveland, Ohio, he migrated with his family to Salt Lake City in the 1870s following the construction of the transcontinental railway and the appeasement of the war between a Mormon militia and federal troops. He bought a small mine, the Old Jordan, and established a steady flow of gold and silver with banks in the East. Over the next years, Liberty Holden incorporated the Old Mining and Milling Company with financial headquarters in New York, which allowed him to expand his operations during the decline of silver prices. In the middle of the 1880s, he gained rights to the waters of the West Jordan Canal, in Bingham Junction. The plan was to build a plant to process the mineral before sending it to the railroad station in Ogden.[15]

By the 1890s, the elder Holden had amassed a fortune of $15 million. Together with R. C. Chambers, owner of the Ontario mine in Park City and a key member of the silver trust, Holden, who was founder and member of the executive committee of the Bimetallic Coinage Association, formed part of a generation of new metal tycoons in the West of the United States.[16] During those years, the feud between the three "Copper Kings" in Montana over the political control of the state had made national news. Economic competition between William A. Clark, Marcus Daly, and F. Augustus Heinze, the directors of the major copper producers of the state, had translated into a dispute over the state capital. The power that the three businessmen displayed shocked the American West and preceded the Standard Oil Company's scandals in the early 1910s: Jerre Murphy described the history of Montana in those years as "a story of modern feudalism."[17] Christopher Connelly noted the economic, political, and symbolic power of these kings in the West and called their involvement in the democratic process a "devil's vote,[18]" and Mark Twain criticized the blurring between political and economic power in the figure of William Clark, by then a US senator. "By his example," Twain remarked, "he has so excused and so sweetened corruption that in Montana it no longer has an offensive smell."[19]

Liberty Holden was not as flamboyant as the Copper Kings, but he was no common businessman. He was a relevant Democrat in Cleveland,

and in 1886, he almost became governor of the territory of Utah. In 1890, he was part of the silver lobby that pushed for the Sherman Silver Purchase Act, forcing the federal government to buy silver as money reserves. He then moved from the silver business into the industrial metals, pursuing the rich copper, lead, and zinc deposits of Bingham Canyon on an industrial scale. These ores built more than one giant corporation, as the result of the consolidation of smaller companies. Simultaneously, Samuel Newhouse organized the Utah Consolidated Company in 1896 as the merging of several mines around the Highland Boy property in the canyon. Three years later, Newhouse sold his interests to the Amalgamated Copper Company, which started the construction of a smelter on the Jordan River, on the edge of the Oquirrh Mountains and within the Salt Lake Valley.[20]

Not only was Utah expanding its economic relevance, but also the mining game was changing by the late 1890s. Corporate officials and technicians had tried to solve the problem of competition through industrial agreements. In 1894, corporate officials of smelters across the country had a conference in Denver, and they formed the Smelters Association. The organization had an objective to reduce the competition for ore and negotiate better freight rates with railway companies. The experiment was short-lived, but it laid out the foundations for the conformation of vast conglomerates in the next decade that integrated the industry and replaced the charismatic leadership of the early tycoons.[21] The dispute between Daly, Clark, and Heinze ended only with a strong dose of corporate consolidation. In 1899, Henry H. Rogers and William Rockefeller absorbed the Anaconda Copper Company of Marcus Daly into a giant holding: the Amalgamated Copper Company. Six years later, the Anaconda reabsorbed most of the holdings of the Amalgamated and acquired the assets of the competitor Heinze, while Clark's company rather rapidly sold their remaining properties.[22] Other titans in the metallic business were born in the same years as a response to competition between big companies, most of them with the same set of actors: technicians. Henry Rogers and Leonard Lewisohn founded the American Smelting, Refining, and Mining Company (ASARCO) as a holding of sixteen metal smelters in 1899, and two years later, the new corporation merged with several properties of the Guggenheim family.[23] The old metallic trusts were falling into the hands of the giant corporate conglomerates. The dynamic of these new behemoths required four elements: a significant market share, capitalization and mergers in the markets of New York and Boston, scaling of operations in processing plants, and consolidation of the mines underground.

Mineralization in the mountains is an irregular process that follows different geological phenomena: fractures, volcanic activity, and crystallization. The concentration of the unique elements in ore varies, and traditional mining followed the richest concentrations of metals in the rock. Following these veins, miners claimed property over the minerals and a path for prospective expansion.[24] However, the concentration of operations in a single district generated disputes over rights and operative inefficiencies underground. A single shaft or drainage could cross dozens of different claims and therefore imply endless negotiations with owners about costs and property of the ores extracted. Only a centralized property could provide the volume of ore required for industrial processing plants; only a large-scale plant could produce metals from low-grade ores profitably; only conglomerated selling firms could control enough the market to protect long-term investments; only large capitalized corporations could invest in acquiring property, developing technology and speculation in the market. Big and bigger became the name of the game.

Liberty Holden eventually retired back to Cleveland and left the business in the hands of his son Albert. Albert had migrated to Salt Lake with his father in the 1870s and then studied at Harvard and MIT. More than an entrepreneur, Albert was a technician and learned by working on the company properties in Bingham Canyon, including the mule tramway business. In 1899, he formed the United States Mining Company. The model of the new firm exploited an interaction between the Boston shares market, the New York metals market, and the western mines. The US Company tried to integrate extraction and processing locally, instead of sending concentrates to smelters in Montana and Arizona. In 1901, the United States Mining Company started the construction of a modern smelter on the Jordan River, at the junction of Bingham Canyon, on a property called Midvale. Two years later, the company completed the installation of an aerial tramway to transport the minerals from the Old Jordan into the Midvale plant, where the scale of the smelters allowed the extraction of copper from low-grade minerals. The rest of the companies chased that innovation. In 1901, the ASARCO inaugurated its Murray Smelter, a few miles from the US Company construction site.[25] In 1905, Victor Clement and Daniel C. Jackling began the construction of a smelter at Garfield, to treat the porphyry copper from the Utah Copper Company. In 1910, the Utah Consolidated was absorbed into the International Smelting Company, building a smelter in Tooele, to the west of the valley.[26]

By the beginning of the twentieth century, the smaller operators in Bingham Canyon were practically extinct, and the few remaining depended on the plants of the four titans: International Smelting Copper (later Anaconda), Utah Copper (later Kennecott), ASARCO, and US Mining Company. In Utah alone, the US Company controlled over 160 claims, formerly from smaller mines. The integration on the surface, the consolidation of the underground, and the capitalization in the Boston and New York markets allowed these firms to expand rapidly. By the end of 1905, the US Mining Company owned most of the shares in the Centennial Eureka Mining Company in Tintic, Utah; a copper refinery in Chrome, New Jersey; the Richmond Mining Company and the Eureka Consolidated Mining Company in Eureka, Nevada; the Mammoth Copper Company in Shasta County, California; and shares in the American Zinc, Lead and Smelting Company, and in the US Oil Company.[27]

These transformations and consolidations in the mines and plants also affected the structure of the financial markets around the firms. Big corporations shared the mining spaces on the ground, in the processing valleys, and among the same financial circuits. The Bankers Trust served as the transfer agent, and the Guaranty Trust as the registrar of stock of the US Company and several of its divisions. This financial duo also managed the shares of the Kennecott Copper Corporation and the Anaconda Copper Mining Corporation, of local banks such as the Salt Lake Security Trust, and other corporations in the area such as the powerful Utah-Idaho Sugar Company. The financial trusts shared most of their directors, and Henry Pomeroy Davison presided over both financial institutions. Simultaneously, Davison was president of the Liberty National Bank, vice president of the First National Bank, the second partner of J. P. Morgan and Co., just after J. P. Morgan himself, and member of the executive board of Chase National Bank. This last institution managed the shares of the remaining member of the copper trust, ASARCO.[28]

The generation of Albert Holden had replaced the old mining tycoons from his fathers' generation, which in turn had grown from the mining rushes in the 1870s to the 1890s. The structure built by Albert Holden allowed the US Company to turn into one of the major players in the metallic business within a single decade. Incorporating the United States Smelting, Refining, and Mining Company in 1906 in Maine was, in fact, one of the last steps of the firm to gain continental relevance. The capitalization achieved through that operation allowed the firm to mutate from a national giant into an international organization. This new structure needed, very soon, the emergence of a corporate technical bureaucracy that would replace the leadership of prominent technicians such as Holden in the following decades. This new machine, the multinational mining corporation, displaced the local technical elites in the South.

PACHUCA, HIDALGO

By 1906, the mining district of Real del Monte and Pachuca had been a dynamic center for mining operations and innovations for four centuries. In 1555, Bartolomé de Medina and Gaspar Loman perfected the method of gold and silver extraction through the patio system in that district. The patio system was a complex process of selection, classification, crushing, separation, treatment with mercury, and smelting of ores that used, as a central reaction, the amalgamation of mercury with oxidized silver and gold ores. The method expanded not only within the Spanish Empire but also worldwide and was the principal method of extraction of gold and silver until the end of the nineteenth century. While Medina's mill in Pachuca formed part of a boom in mining in the district by relatively small companies, the rush ended in the seventeenth century. A new boom in the area started in the second third of the eighteenth, this time based on economies of extraction underground. In 1738, José Alejandro de Bustamante and Pedro Romero de Terreros started a company to exploit a vein in Real del Monte, the *veta Vizcaína.* The company was a classical colonial firm, with access to indigenous work supported by the crown and organized in twenty-four *barras,* or shares. The company's innovation was to make a tunnel, a *socavón,* to drain the several shafts of the district of Real del Monte. Even after Bustamante's death in 1750, the tunnel worked and initiated a second bonanza in the district under the leadership of Romero de Terreros.[29]

By the beginning of the nineteenth century, the boom was over, and the successors of Romero de Terreros, now the Counts of Regla, could not continue the infrastructure requirements to remain profitable. Much as in Bingham a hundred years later, extensive construction projects underground implied negotiations over rights to ores that could be expensive to litigate. The independence movement interrupted the project of a second drainage tunnel in the district in 1819, and the third Count Regla looked for external sources of capital. In 1823, he established an association with English agents, forming the Compañía de los Aventureros de las Minas de Real del Monte. The new firm was a typical free-standing company that separated management in Mexico with ownership in London. Access to capital allowed some development work in Real del Monte, using Cornish technology to mechanize the water drainage. Like many other companies formed during the British speculative investment boom to the Americas, management refused to incorporate local expertise and encountered major obstacles. In particular, their obsession with draining the veta Vizcaína resulted in continuous losses to shareholders. After twenty-five years, and without a return on their investment, the shareholders sold the company to Mexican investors in 1849 at under 10 percent of their book value.[30]

The new Mexican owners maintained the model of the firm by not intervening in the administration and keeping the Cornish management. Less concerned with the legendary Vizcaína vein, the new owners allowed engineers to exploit the low-grade ores of Pachuca. Just four years later, the Compañía hit a bonanza that lasted thirty years. In the following decades, they modernized the amalgamation plants, incorporating mechanized processes, and the company pressed for a coinage facility in the district. By the 1880s, the company had introduced access to the railroad and the use of electricity into their Loreto mill. However, as with the mines in Bingham at that time, the depression of the silver prices pushed the company to restrict their operations to high-grade ores, incapable of profiting from the efficiency in the plants. In the last years of the nineteenth century, they started experiments with the ores with a new technology, the MacArthur-Forrest process of cyanidation, used successfully in gold ores in 1890 in Witwatersrand, South Africa, and the Mercur mine in Utah, in 1891.[31] The first experiments were unsuccessful, and although the Compañía had powerful incentives, it lacked the capital necessary for productive renovation.

Big American corporations were by then pushing capital outside the United States. In the last years of the nineteenth century and the beginning of the twentieth century, American companies purchased firms in the historic mining districts in Mexico. In 1890, Samuel Newhouse and the Guggenheim brothers negotiated with the Mexican dictator, Porfirio Díaz, regarding the establishment of two smelter plants in Mexico. One would be in Aguascalientes, processing the silver, gold, copper, and lead metals in the center of the country, and the other in Monterrey, treating the ferrous metals from the northeast.[32] Soon, other big mining corporations followed and established big companies and plants shipping metal to the New York market. In 1906, the Amalgamated Copper Company purchased the Cananea Consolidated Copper Company on the northern border from Colonel William C. Greene, taking total control of the company in February of the next year.[33]

But the problem of silver cyanidation with Mexican minerals still had to be solved. New, small-scale American firms had succeeded in using it with gold-bearing ores in Sinaloa, Sonora, and Chihuahua, and by 1902, local technicians in Mexico City could replicate the MacArthur process. As a result, some miners reported limited extraction of silver, as the global price of the metal increased. In 1905, two events marked the acceleration of investment into Mexican silver mines. First, in February, the Guanajuato Consolidated Mining and Milling Company reported that its new cyanidation plant for silver-bearing ores was complete and successful. Second, the Mexican government abandoned bimetallism and adopted the gold standard in March and April, which liberated the silver price.[34]

The management of the US Company had been looking for investment opportunities in Mexico. The firm hired Pablo Martínez del Rio, the director of the Central Mexico Railway and from one of the most prominent bourgeois families in the country, to lobby for the establishment of a smelter in Hidalgo del Parral, Chihuahua, close to El Paso. From May to July 1905, Martínez del Rio asked the minister of the treasury, José Yves Limantour, and the governor of Chihuahua, Enrique Creel, but they informed him that Guggenheim and the American Smelting and Refining Company had the privilege to install a plant in that area. Martínez del Rio had a second option for the firm. His family had been part of the Mexican investor group of the Compañía Real del Monte since the 1850s, and Felix Cuevas, president of the executive board, had been struggling to maintain the profits of the firm. As the Compañía was incorporated in London, the proposal to buy went through the intermediation of the Venture Corporation Limited, based in England.[35]

The London Venture Corporation had been the instrument of expansion of British capital and US engineers in Mexico in previous years. The firm had been the intermediary for the Guggenheims, advised by John Hays Hammond, to secure the control of the Esperanza mine in the state of Mexico and invest in the region of Minas Prietas, Sonora. This time, the Venture Corporation had to transfer the control and capital of the Compañía de Real del Monte from London to Boston. In November, Venture Corporation. acquired 51 percent of the shares, and on January 9, 1906, the US Smelting, Refining, and Mining Company was incorporated in Maine in order to raise capital and invest in Mexico. The shares were traded in Boston, and by the spring of next year, the US Company held 100 percent of the shares of the historic company in Hidalgo.[36]

Now the parent company, The US Company reorganized the local firm immediately. They kept prominent Mexican politicians as formal directors: Nicolas Martinez del Rio, Julio Limantour, and Luis Riba.[37] Pablo Martínez del Río visited the governor of the state, Pedro L. Rodríguez, with an introduction from José Yves Limantour himself. In the letter, the secretary of the treasury explained the changes in the company and supported the new executive committee of the firm, represented by Pablo Martínez del Rio and Luis Elguero, as well as the new general manager of the subsidiary, Morrill B. Spaulding.[38] On May 15, 1906, the officials of the firm pushed for the foundation of the first mining chamber in the country, along with Rodolfo Reyes, Oscar Braniff, Edmundo Girault, and Gabriel Mancera, while Luis Elguero took part in the director's board. Pablo Martínez del Rio became the president of the new organization, and two weeks later Porfirio Díaz welcomed the new organization into the presidential palace.[39]

To that point, it appeared as if the old elite had managed once again to maintain control of the silver in Pachuca, but it was an illusion. The new management rapidly transformed the silver mining business in Mexico, mainly concerned with technical changes. Spaulding, an engineer from New York, had experience with the cyanidation process in the San Sebastián area in El Salvador. He was in charge of the technical crew of the new division, but he was also appointed to the director's board of the mining chamber.[40]

The Compañía had mutated from a family-controlled firm to a British stand-alone and then to an independent Mexican company. Each period reflected a particular integration of Mexico, and silver, into the world market and the structure of global imperialism. Beginning in 1906, the old Compañía de Real del Monte y Pachuca would become one constituent part of an Underground Leviathan, a multidivisional firm that satisfied the North American hunger for metals. Over the next decade, and even after the explosion of the Mexican Revolution in 1910, American firms in Mexico grew exponentially. ASARCO and Phelps Dodge increased their investments in the northern states of Chihuahua, Coahuila, and Sonora. At the same time, the US Company cemented itself as the largest mining player in the country. In 1908, it finished the first cyanidation plant in Real del Monte, the Guerrero mill, and appointed Carlos W. Van Law as the new director of that division. Van Law was one pioneer in the cyanidation methods in central Mexico and had represented the Guanajuato Reduction and Mining Company in the mining chamber.[41] Over the next five years, the company multiplied its production fivefold. Technicians as Van Law and Spaulding took control over all the decisions of the Mexican division, and little by little replaced all the Mexican elite from management positions.

While the Mexican Revolution impacted the operations of the firm, it was the war in Europe which temporarily slowed down the growth, as explosives and cyanide became scarce. The company stopped operations only in the month following the US invasion of Veracruz in 1914, and after the punitive expedition to find Pancho Villa in 1916. On average, during the deadliest war in Mexican history, the Compañía de Real del Monte operated at between 50 percent and 80 percent of its maximum capacity.[42] With the cessation of hostilities, by the beginning of the 1920s, the investment of US firms in Mexican mines was three times as big as in 1900, and 25 percent larger than in 1911.[43]

Like what had happened in Utah, the expansion of the US Company in Mexico not only relied on the economies of the plant but also in its growing control over the mining claims in the district. Integration on the surface depended on the same mechanisms as in Bingham. The firm installed three lines of an aerial tramway from Real del Monte, Atotonilco El Chico, and Santa Ana to the Loreto mill in Pachuca while centralizing underground the different mines to the north, northeast, and northwest of the plant. By the 1930s, the Loreto mill could separate and treat profitably not only silver and gold, but also zinc and lead-bearing ores, and had installed an electrolytic refinery, allowing the corporation to export refined bullion. The growth in the processing capabilities increased the centralization of milling and refining activities in the area around the US Company, establishing long-standing agreements with small and medium-sized companies across the state of Hidalgo.[44]

Scale economies in the plant fostered the consolidation of the mines in the district around the US Company. There are five major groups of metallic veins in the mining district of Pachuca: the Vizcaína, El Cristo, San Juan Analco, Santa Gertrudis, and Polo Norte. These fracture systems extended across over two hundred different mines claimed in the three hundred years of history of mining in the district. By 1916, the US Company had concentrated 65 claims in Real del Monte and 119 in Pachuca, for 2,000 acres.[45] Most of the mines that the company did not own were actually under control of the firm, as it was the only one that could provide the infrastructure to make profitable the extraction of minerals. Technically, centralization allowed the firm to better drain the different properties, decrease the number of shafts, and reduce costs of transportation on the surface. The centralization of the underground required communicating between three different districts under the surface, with over five thousand miles of tunnels. This invisible city absorbed some of the largest companies operating in the district: Las Maravillas, San Rafael y Anexas, Compañía General de Invenciones Mineras, and Compañía Explotadora de Minas. By the 1930s, when the project of underground centralization concluded, only the Santa Gertrudis Company and the Dos Carlos Company operated independently of the American giant in Hidalgo. Along with the US Company, these these two firms, which were among the biggest British stand-alone firms in the world, appeared as secondary and dependent local enterprises.[46]

EAST CHICAGO, INDIANA

The expansion of the corporate body on the continent mediated the relationship between the management headquarters in Boston and the metallic markets in New York. Coordination from management in Boston required an ever-growing bureaucratic apparatus and an efficient mechanism for international coordination and communication. The absence of any strong shareholder presence implied that the governance structure of the multinational depended on the design of rules by directors and executives. The corporation was a self-regulated organism composed of technicians, engineers, geologists, managers, accountants, lawyers, metallurgists, and chemists, among others. In this social system, orders and power were distributed top-down, but reports and information circulated from the bottom up. The foremen reported on operations to the superintendents, who elaborated more extensive accounts to general managers, who, in turn, made a weekly abstract to the parent company officials. These dispatches allowed directors to make informed investment decisions, containing information from all levels of the organization, from the exploration of new properties to the existence of union organizers among miners. Nonetheless, the firm had additional methods of controlling the day-to-day operations. The auditing department verified the economic information given by the reports, while the general officers in Boston oversaw the control of the major technical decisions. The auditor's office, independent from the auditing department, conducted regular evaluations of the performance of subsidiaries, besides having randomized in-depth examinations of them. The headquarters in Boston had a general engineer, metallurgist, mechanical engineer, geologist, and chemist, who tested the major productive decisions made by middle officials in subsidiaries.[47]

With a center in Boston, this bureaucratic apparatus was the real authority inside the Leviathan, which would expand to the West of the United States, central Mexico, and the industrial Midwest. Power was exercised from Boston out, but the metals traveled from the continent into the New York port. In the case of the US Company, the firm responsible for distributing the market operations in New York was the International Metals Selling Company, which specialized in the commercialization of the concentrates, bullions, and refined metals of the distinct divisions of the US Company. This subsidiary in New York analyzed prices and made projections to maximize the revenues obtained by the sale of metals, consulting the divisions on the best moments to ship their products. The local firms of the multinational corporation sold their output to the International Metals Selling Company on a competitive basis with the price published in the *Engineering and Mining Journal*. The International, on its side, obtained an intermediation fee but would decide the moment of sale. The system of the subsidiary was not unique, as one division of the Anaconda Copper Company, the United Metals Selling Company, controlled over half of the market of copper at the time. In the next years, the subsidiary of the US Company became a fully integrated department of the mining multinational.[48]

Expansion and internationalization of these mining corporations also altered the industrial structure of the East and Midwest of the United States. The first refinery of the Guggenheim Exploration Company, later absorbed into the ASARCO properties, was built in Perth Amboy, New Jersey, in 1894. Its primary objective was to treat the concentrates from the company's smelter in Aguascalientes, which traveled by railway to the Tampico port, in Tamaulipas, and then by ship to the East Coast.[49] The model was replicated by the United States Metals Refining Company, a firm that integrated into the US Company conglomerate in 1904 and that owned a plant in Chrome, New Jersey. Following negotiations for the purchase of a Mexican division, the US Company started planning to install a lead refinery in East Chicago, Indiana.[50] Other firms followed as the southwest of the Michigan lake was transformed into a metal manufacturing cluster. In 1908, the Anaconda Copper acquired a refinery in Perth Amboy and started to build a lead refinery in East Chicago, Indiana. The new plant was said to be the largest lead producer in the world.[51] By 1912, the American Smelting, Refining, and Mining Company had also acquired a refinery south of Chicago, which became its national plant.[52]

The metal refineries in East Chicago reflected the changing nature of the metal market and the plasticity of the metallic mining corporations to adapt to the ever-evolving prices and concentrations in their ores. Not all ores are created the same, and huge disparities arose between the concentrates obtained from the multiple divisions and mines of the firms. Besides, certain waste compounds could become, suddenly, subproducts if the prices were correct. The refinery of the US Company in East Chicago reflected many of the transformations of both the territorial expansion of the firm and the nature of mineral prices in the global market, articulating materially the products from the smelter in the south and west into commercial compounds in New York, from gold and silver to white lead, refined copper, or sodium arsenite used as a pesticide.[53]

The economies of scale of these large multinationals kept on growing in the next decades, and they started to internalize not only the extraction, transformation, refining, and selling of minerals, but also to control most of their supplies. Smelters of big mining corporations usually depended on small subsidiaries of the parent company for their supplies of coal, useful in the high furnace process. The US Fuel Company of the US Company, the coal department of Anaconda Copper, and the Peabody Coal Company of Kennecott all supplied the heat that transformed the metal. Lime and timber departments of subsidiaries were also present in Montana and Utah, supplying the mines and the smelters, and could sell some of the same inputs to smaller companies nearby.[54]

In short, the tremendous growth of the mining multinationals in the beginning of the twentieth century was not only the result of expansive tendencies inside the bureaucratic apparatus of firms. In effect, most of the new additions to the firm, as with the Mexican case, implied smaller companies who were willing to be part of a multinational organism presenting acquisition opportunities to bigger firms. After the acquisition by a big multinational, managers and engineers of smaller companies usually stayed as corporate officials in similar hierarchies. By merging with a corporate giant, small companies could replace the economic competition in the market with technical negotiations inside the firm, and could access capital and expertise from the headquarters. The multinational corporation was not born on its own, but it was instead called into being.[55]

Scale was an important drive to the growth of the mining firms in the beginning of the twentieth century, but it was boosted by the inherent nature of mining. A metallic ore takes millions of years to make, and most corporations cannot wait that much time to obtain profits. In consequence, a territory, once exploited, transforms from valuable metal rock into waste. Luckily, the amount of materials used in mining is an insignificant fraction of the ones available on the crust of the planet. This produces an odd result: mining is essentially unsustainable but practically unlimited. In other words, mining requires an endless growth of its boundaries, and mining firms are always a frontier enterprise. The firm had a special mechanism to keep absorbing new properties into their system of expansion: the exploration department.

The Underground Leviathan requires aggressive incorporation of matter into its system, and the subsidiaries of the corporation were not the only ones that satisfied that hunger. In the first years of the twentieth century, mining multinationals designed subsidiaries that specialized in increasing the frontier of the corporation: locating, evaluating, and developing new mines. These exploration divisions translated the non-territorial sovereignty of the US Company into a spatial reality in the local area.

The Guggenheim Exploration Company was the main referent for the establishment of an exploration division in the metallic behemoths. Guggenex had been responsible for buying small mines in Mexico after ASARCO installed the smelters in Aguascalientes and Monterrey in the last decade of the nineteenth century. It was Guggenex that, in 1905, purchased most of the shares of the Utah Copper Company from Victor Clement and Daniel Jackling and pushed for the construction of the Garfield Smelter in the Salt Lake. In 1909, Guggenex reorganized the Braden Copper Company in Delaware, in order to develop the El Teniente mine in the north of Chile. In 1912, the firm organized the Chile Exploration Company in New Jersey, later reorganized in Maine, to take over the possessions of Albert C. Burrage in the territory of Chuquicamata, also in northern Chile.[56] In short, Guggenex was the main vehicle for the growth of the metallic empire of the Guggenheim in the first years of the twentieth century.

The US Exploration Company performed this role for the US Company. Following the acquisition of the Mexican firm in 1907, the US Company subsidiary in Mexico started the operation of the new cyanide mill and, with it, the extraction of ore in the mines of Pachuca. That same year, the firm invested in a new corporation: the Peruvian Mining, Smelting and Refining Company. The firm was created in the United States to obtain capital for the exploitation of the Churraca mine in the Morococha district. The US Company became, along with the Cerro de Pasco Mining Company, one of the most massive corporations investing in the old silver mines in Yauli, Peru, after successful experiments of extraction of copper in the ores. As in the Mexican and Chilean cases, this first influx of US capital took over old claims and firms, but the rapid growth of the mining corporation soon reached an operative limit that required a new institutionalization. In 1908, Sidney J. Jennings, the engineer in charge of the evaluation of the Peruvian properties for the US Company, was elected vice president of exploration and mining investment within the holding company. The step allowed the firm to make a permanent department in charge of surveying activities in Latin America. Simultaneously, the US Company created the Department of Structural Geology under the command of W. F. Ferrier, a member of the Canadian Geological Survey.[57]

Both Ferrier and Jennings had experience as surveyors of Anglo-American interests in the world, advancing the mineral frontier of the North Atlantic on different levels. Sidney Jennings was originally from Kentucky but studied in France and Germany before getting a degree from the Lawrence Scientific School at Harvard in 1885. His brother Hennen had previous experience in mining in California and Venezuela, before he and Sidney traveled to South Africa in the last decades of the nineteenth century. There, Hennen worked for the Wernher, Beit & Co., while Sidney became manager of the Willows Copper Argentiferous Syndicate, at the De Boers Consolidated Mines, Crown Deep Ltd., Crown Reer Gold Mining Co. and a consulting engineer at the Robinson Gold Mining Co., H. Eckstein & Co. Soon Sidney became one of the most prominent engineers in the territory, and was elected president of the South African Association of Engineers and vice president of the South African Association for the Advancement of Science. In 1907, the Jennings brothers came back to America. Hennen kept a consulting business in Washington, while Sidney accepted the offer of the US Company to evaluate properties in South America.[58]

W. F. Ferrier was an expert in the operations of the Pacific coast and in the north of the continent. Born in Rossland at the end of the nineteenth century he produced a complete survey of the mineral potentialities of British Columbia, collecting and curating the rocks used in the Canadian exhibit for the World's Columbian Exposition in Chicago. In 1898, he worked as a surveyor in Mexico, and in 1907, he started working for the US Company in the Mammoth division in Shasta County, California. His new department had the task of producing geological studies of all the parent firm's prospective properties. As head of the structural geology department, Ferrier examined properties for the firm in Idaho, where he found the first fossil of an American Helicoprion, and in the Portland Canal, British Columbia, while remaining based in California.[59]

In 1912, the structural geology department and the exploration of mining investment department merged into a new subsidiary. The resulting United States Exploration Company established three offices: one under Ferrier in Seattle, for the evaluation of properties in British Columbia and Alaska; one in the headquarters in Boston, with Jennings in charge; and one in Mexico City. The US Smelting Exploration Company of Mexico, SA, was chartered in Mexico, and its function was to serve as a pivot between the US Exploration Company and the Mexican subsidiary, the Compañía de Real del Monte y Pachuca. This division of the exploration department employed almost exclusively geologists and technicians of the Mexican division. In the next years, the company, as with the rest of the operatives, moved its offices to Pachuca and founded another firm, the Compañía Exploradora de Hidalgo, SA, which specialized in small claims.[60]

In particular, the growth of the US Exploration Company preyed on the decline of the British exploration syndicates, which operated as stand-alone companies but depended, even more than traditional British firms, on local alliances. The Mexican division of the exploration department had most of its claims, up to 80 percent of them, in the Hidalgo vicinity, absorbing most of the territory of smaller British and American firms. The British-Mexican Exploration Syndicate, which controlled the Dos Carlos mines, and the Camp Bird Limited, which owned the Santa Gertrudis mines, were important and profitable firms. But the companies, based in London, could not match the scale of the American mining corporation, even with good access to capital and technology. Both of them had to sell an increasing number of properties to the US Exploration Company, and some engineers of these firms had to move on into other ventures or sectors. This was the case of the chairman of the Camp Bird Limited, Herbert Hoover.[61]

As with the new divisions, most of the absorption of territory started by the proprietors of valuable claims who submitted their prospects to the exploration subsidiary. In 1911 alone, the exploration division received 921 purchase options, rejected 749 immediately, and refused

144 after preliminary revision, examining 28 altogether and taking only one option, the properties of the Consolidated Fuel Company in southern Utah. In 1912, the Seattle office of the US Exploration Company acquired all the shares of the Consolidated Fuel Company, along with other firms in the region: the Castle Valley Coal Company, the Black Hawk Company, and the Southern Utah Railroad. The consolidation of these smaller firms allowed the exploration division to rationalize the transportation of the bituminous coal and improve the communications among them. In 1913, the holding created a new firm, the US Fuel Company, and the Utah Railroad that managed the operations of the railroad.[62]

Most of this expansion was a gradual process of monopolization of the mining spaces across the country, through the acquisition of smaller, American-managed firms. The US Smelting Exploration Company in Mexico developed two different interests: the expansion into traditional silver and zinc districts in southern Mexico and beyond, and the exploration of copper and iron properties in the North. The firm invested in silver and gold mines in Xito, in the state of Guanajuato; Xitinga, in the state of Guerrero; and Real de Catorce, in the state of San Luis Potosi. All three were traditional mining properties that could only be profitable with the scale technologies of the firm and the use of its circulation network. After the fall of Porfirio Díaz, the US Company could expand its activities in the north and the west. In the north, the exploration division purchased the Artemisa Group, close to Hermosillo, Sonora, from the Tecolote Copper Corporation of Arizona. The firm could, after the fall in grace of the old elites allied with the Guggenheim interests, also expand its properties in Chihuahua. The divisions installed a new mill in Ciudad Jimenez, Chihuahua, which served the copper minerals from many merged properties in Hidalgo del Parral. In the next years, the firm acquired claims to twelve different companies in the area and extending to Batopilas, Ojinaga, and Santa Barbara. The company also purchased mines in the neighboring state of Durango to send to Parral, in Guanacevi, Santa María del Oro, the Mala Noche Group, and San Fernando. At the same time, the US Exploration Company secured properties of the Sonora Placer Company in Sinaloa, Sonora, and Chihuahua; the Magistral Group in Jalisco; coal mines in Coahuila; and the Blanca mines in Agualeguas, Nuevo León, from ASARCO. By 1935, after its first investment in Mexico, the US Exploration Company owned over 13,000 mining claims in nine states of Mexico.[63]

The establishment of the US Exploration Company office in Mexico reveals the role of the Mexican subsidiary in the projected expansion of the firm and the general pivotal role of Mexico in the expansion of US influence in the continent. During those years, American companies enjoyed extremely favorable conditions for development in the area. "There are no taxes of any kind on mining or milling lands, with free entry of all kinds of necessary machinery and appliances through the custom-houses of the Republic," explained the International Bureau of the American Republics about the benefits of doing business in Central America.[64]

Expansion to the mineralized territories of Central America soon became a long-standing project of the firm. Morrill B. Spaulding, the first general manager of the Mexican division, had worked in the gold mines of El Salvador in the first years of the century and knew of possibilities for small developments. In 1912, the US Exploration Company office in Mexico sent William Davis to examine some properties in Yuscaran, Honduras. During the next two decades, the firm continuously sent new engineers to Tegucigalpa. In 1932, the division invested heavily in surveying properties around Danli, to the east of Tegucigalpa, and close to the border with Nicaragua, a mineralized region with sulfide deposits of gold and silver. In the next years, the firm analyzed samples of the gold and silver sulfides in the areas of Danli, Yuscaran, and El Tránsito, the latter closer to the border with El Salvador. The main officers of the Mexican division, C. B. E. Douglas and M. H. Kuryla, traveled to the area, evaluating the possibility of integrating it into the operation in Mexico, budgeting the cost of aerial transportation and the railway routes from Tegucigalpa to Pachuca.[65]

The Peruvian options that Jennings had evaluated earlier were, by then, abandoned by the firm, but after the excitement of the Honduras project, the Mexican division of the US Exploration Company contemplated an extension to the mining districts in Colombia and Venezuela. In the summer of 1932, the firm received samples of molybdenum from Nicolas F. de Roses, as well as general information on promising opportunities around Barranquilla, Colombia. The firm extended the surveying of properties in the district of Santander, Colombia, but some of the most promising fields were in Venezuela. Between 1930 and 1933, the US Exploration Company invested in the analysis of the economic potentialities of the gold and silver dredging camps in the southwest of Venezuela, to the south of the Orinoco.[66] Placer mining in the south was not new for the firm. Over the previous decade, the office in Seattle had invested in an option that became central in the next years of the holding: a massive placer mining operation in Alaska.

FAIRBANKS, ALASKA

The waves of rushers to Alaska, thirty years after the acquisition of the territory from Russia, replicated some of the social and symbolic characteristics of earlier migrations to the American West. Individuals searched for fortune in surveying lands, panning rivers, hunting for fur, and fishing on the coast. White settlers established and betrayed alliances with the native population, and legalized and regulated prostitution. Cities exploded and collapsed in a matter of years, and the newcomers enjoyed a free but brutal experience in their quest to control the wild. This time, though, corporations formed part of this expansion from nearly the beginning. Successful individuals did not seek to become owners of mines, just their sellers. Surveying and panning were the beginning of a process that, if successful, would end with a sale to a large-scale firm that could make exploitation profitable.[67]

The biggest player in the territory emerged at the turn of the century. In 1900, an indigenous expert guided Clarence Warner and Jack Smith to a copper cliff to the east of the Kennicott Glacier. The deposit had exposed copper of 70 percent of purity, and the claim change renamed the space as the Bonanza Mine Outcrop. When Stephen Birch heard of the discovery, he started to acquire options in the slope, and, like every other entrepreneur in the mining sector, quickly searched for buyers. With the involvement of the Havemeyer family, J. P. Morgan, and Daniel Guggenheim, Birch organized the Alaska Syndicate Company, which developed the mines. The Syndicate started to extract copper in 1911 and transformed into the Kennecott Copper Corporation four years later. The new corporation became so powerful that it absorbed some other of the Guggenheim properties, such as the El Teniente and Chuquicamata mines and the Utah Copper Company.[68]

Kennecott was by far the largest player in the farthest north territory of America, but it was not the only one rolling the dice. John Hays Hammond, the Midas King of the Guggenheims, began dredges to smaller miners in the territory and analyzed other good options. In 1912, he founded the Yukon Gold Company, a property of Guggenex, and consolidated several placer mines in interior Alaska, close to the Kenai River. The US Company was another of the big players that tried to operate in the new frontier. In 1911, the firm conducted technical surveys of different properties and two years later developed the Ebner mine in Juneau. The open-pit property began two decades before, by William M. Ebner, and had two shafts and a small separation plant. After seven years of improvements, the US Company concluded that it was impossible to make profitable the exploitation of the sulfides in the mine at current prices, and abandoned the works.[69]

Despite the failure, the firm continued to survey in the area, but the opportunity to install a division came years later, at the hand of Wendell P. Hammon. This mining engineer from California had started, like John Hays Hammond, selling small dredges to rushers in 1907. The replication of the technology used in the West Coast soon increased his attention and, three years later, the Hammon Engineering Company sent Newton Cleveland to install and operate a dredge on the Solomon River in the Seward Peninsula. At the same time, Hammon joined Daniel C. Jackling, the engineer who founded the Utah Copper Mining Company, in the development of the Alaska-Gastineau mine in Juneau. As with most of the operations in Juneau, the hard rock mine was not particularly profitable, especially adding the high transportation costs. Large-scale dredging could only be implemented near the coast, and the miners in the interior used steam to thaw the permafrost when no other sources of water were available. In Nome, 1919, Hammon tried a new technique, using cold high-pressure water to melt the hard permafrost. The innovation was a success that revolutionized gold dredging in the rivers of Alaska for the rest of the century.[70]

In 1922, the same engineer founded the Hammon Consolidated Gold Fields Company in Nome, the closest urban population to Russia in the Seward Peninsula. He acquired most of the properties in the city from the old Alaska Mines Corporation and the Miami Company, and two big dredges from the Yuba Manufacturing Company. During the winter of that year, he negotiated with the US Company for its participation in the next season. It was a good fit for the exploration department. In 1923, the investment of the metallic corporation doubled the size of the Hammon Consolidated Gold Fields Company and started to scout the possibilities in interior Alaska. In the summer of 1924, Hammon founded, exclusively with US Company capital, the Fairbanks Exploration Company. As in Nome, they purchased most of the small placer properties of old rushers. After just one year of operation, the firm decided that the innovator was not a good planner, and they decided to buy all his shares in the business. The company took the name of the exploration department, but soon became a fully fleshed subsidiary inside the corporate behemoth. The new subsidiary expanded its operations in the peninsula, installing new infrastructure for the mining camps. . By 1928, the company employed, either in the dredges or construction, 80 percent of the population of the town, and it became the core of the towns of Fairbanks and Nome and the biggest mining firm in the territory.[71] The new division, the northern mining operation in the Americas, was a long-ambitioned member of the continental sovereignty of the US Company.

From a relatively small number of claims in Utah to the gold fields in Alaska, the United States Company designed by Albert Holden had increased its dominium over minerals across the continent, from extraction to the manufacture into end products. Albert Holden died in 1913, before the firm reached this expansion, but technical-minded people like him succeeded in reproducing the model during the next decades. Holden's generation was followed by an even more massive wave of technical experts that extended the power of the mining behemoths in the Americas. On the verge of the Great Depression, five companies dominated the nonferrous metals in the entire continent, from the Bering Strait to the Chilean desert. The US Company operated the largest silver mine in the world, in Mexico, and the largest gold deposits in Alaska; Kennecott Corporation owned the El Teniente mine and the Kennicott deposits; ASARCO controlled dozens of smelters in North America; Phelps Dodge Corporation had metal mines from Arizona to Peru; and Anaconda controlled copper production in Montana and owned the Chuquicamata mine. Besides these highly profitable properties, all these companies also owned thousands of claims around the hemisphere. The mountains of the continent fed the metallic titans in those years, and they continued to grow as sovereign entities. The organizational machine of the firm was controlled by the visible hands of its officers. They pulled the invisible strings of the gigantic automaton, adjusting the gears of the multiple moving parts. And, although the function of the firm was primarily economic, the minerals and their exploitation were only the foundation for a much more complex continental Leviathan.

In other words, the definition of the United States Company's territorial boundaries was not only an economic definition of limits of the firm, as it was a decision between management and market-directed operations.[72] Neither was it only the Hobbesian political body's definition, establishing a frontier between the natural state of humans, of perpetual war, and civil society under a common power.[73] Finally, the firm's borders reflected not just the expansion of capitalist power in the territory or original accumulation in Marxist terms.[74] This territorial expansion was also the definition of corporate sovereignty on its multiple bodies, economic, political, and material, as well as maintaining the organization and power inside the Leviathan.

The specific form of this artificial animal was the US Company incorporated in Maine, but that was only one of its bodies. The engineers and managers of the firm controlled the basic organization mechanisms of the corporation, but the Leviathan also required a body of subjects, the object of the exercise of Power, and a material body, the tangible object that flowed throughout the business structure. This book will examine some of the firm's different bodies in the next chapters. As they remain underground, for the historian and the people outside the US Company, the multinational corporation had only a ghostly existence. Yet the divisions appear as a material reality, often transforming the local histories of towns in discernible ways. The subsidiaries altered the beach of Nome, facing Russia; built an invisible city in Pachuca, Hidalgo; contributed to digging the biggest hole on Earth in Utah; speculated with silver in the mountains in Bolivia; looked for gold in the Venezuelan jungle; and erected and abandoned countless mines and towns. If the corporation ever had a soul, it navigated the rivers of metal and minerals in the underground in the continent, answering to all the inquirers: "It makes no difference to me where I am."

CHAPTER TWO

The Body of the Subjects

The King has two Capacities, for he has two Bodies, the one whereof is a Body natural, consisting of natural Members as every other Man has, and in this he is subject to Passions and Death as other Men are; the other is a Body politic, and the Members thereof are his Subjects, and he and his Subjects together compose the Corporation, and he is incorporated with them, and they with him.

—*Willion vs. Berkley,* quoted in Kantorowitz in *The King's Two Bodies.*

CORNWALL, ENGLAND

On January 5, 1920, Sidney J. Jennings, chief of the exploration department of the US Company, gave a speech on the labor issue at the Boston section of the American Institute of Mining and Metallurgical Engineers (AIME). He was an accomplished mining engineer, who had worked in South Africa and led the Latin American expansion of the US Company. He had also presided the AIME before and now was on a panel with John F. Perkins, a former member of the US War Labor Board. Jennings compared the situation of labor unrest to an anecdote from the Roman Republic as reported by Livy. The plebs, Jennings narrated, refused to work longer hours for the patricians and left the city en masse. In the next months, plebeians and patricians suffered from shortages of necessary products, and the patrician leaders addressed the plebs with a parable of the human body. In Jennings's meta parable, the hands started a strike against the stomach and the feet, and the brain, the eyes, and the mouth quickly followed, so the entire body became idle. The stomach grew weaker for the lack of food, but weaker still became the rest of the members: "Since the stomach could exist without most of the members, but the members could not exist without the stomach."[1] Jennings' comparison of the workers as members of the human body followed a long-standing tradition of identification between the sovereign and the subjects, but was also firmly based on the identification of workers as subjects in the United States Company. The patricians were the managers, the plebeians were the workers, and the Roman Republic was the multinational corporation.

In the previous decades, the expansion of the US Company into the Intermountain West and central Mexico had transformed the functions and characteristics of the body of subjects of the firm: the workers. More than a part of the corporate body, the workers were one of the multiple bodies of the corporate sovereign with internal logic and self-organization. Both as political and physical bodies, miners reclaimed autonomy in the workspace, while the organizational body of the firm reclaimed control over it. Their political and physical bodies also preceded the formation and operation of the corporate body of the United States Company. When the company started operations in central Mexico and the Intermountain West, workforces were very different. In Pachuca and Real del Monte, in Hidalgo, the corporation faced a mining tradition of over four hundred years. The intensive exploitation of ores in the area in Hidalgo created an early industrial and capitalist organization of labor relations. Laborers in owners' mines, and not rushers, made up most of the miners in the area since the seventeenth century, but workers organized the work in the underground in teams led by a *barretero*. This was a specialized miner who coordinated the direction of the tunnels, the use of explosives, and the removal of the rocks. Payments in the mines since the sixteenth century consisted of two parts: a base wage for laborers, for the daily infrastructure work, and the *partido*, a division of the extracted rock between the crews, led by the barretero, and the company official. Another specialized laborer, the *maestro tortero*, coordinated the work on the surface. He decided the concentrations of mercury, salt, water, and crushed ore for separating the silver and gold from the sulfur and silicates in the rocks and the different times of processing for ores with different concentrations of metals. This basic structure of work had remained basically unaltered until the beginning of the twentieth century when the US Company bought the Compañía de Real del Monte y Pachuca.[2]

This work structure had determined, since the seventeenth century, a strategic position of the *barreteros* in conflicts between labor and capital. In 1765, the barreteros had organized the continent's first industrial strike as an answer to a drop in the workers' remunerations by the Company of Pedro Romero de Terreros. The barretero had the attribution of picking the ore that the crew was able to keep as remuneration. Romero de Terreros, in search of further funds for the construction of a new drainage tunnel in the area, changed the system, dropping the daily wages and mixing the ore before dividing in the partido. The reform fueled a rebellion among the workers, led by the barreteros. Soon, Romero de Terreros fled to Mexico City, fearing for his life, and the company was forced to increase the remunerations of the underground crews.[3]

The British company did not change the structure of wages in the nineteenth century. The Cornish management of the Compañía de Real del Monte y Pachuca tried to get rid of the partido system and replace it with a flat rate, which also implied a more significant control of the engineers and managers working underground. In 1827, 1833, and 1847, the stand-alone British firm tried to modify the partido. The attempts faced a fierce opposition from Mexican workers and determined the alienation and eventual expulsion of the Cornish skilled workers who had migrated with the managers. As the firm still relied on methods of selective mining led by the expert knowledge of the barreteros, the traditional payment system never disappeared under the Cornish management, outlasting English capital and remaining in place under the Mexican owners between 1847 and 1906.[4]

Management's inability to replace the system was not exclusive to Mexican mining and was a widespread phenomenon in both Mexico and England. The Cornish managers and engineers were familiar with the *tut and tribute* system of remuneration in Cornish mines. The *tribute* was the payment of a fraction of the mineral extracted to the worker, or crews of workers, once the owner's costs for infrastructure were deducted. In contrast, the *tut* was the payment of a flat rate by dimension or weight of the extracted rock. Usually, mines combined the two types of remuneration for different types of labor in the mine. Development work, which consisted of the extraction of non-mineralized rock and the reinforcement of the structure, was paid by the tut. In contrast, the profitable work, the extraction of metallic ores, was paid by the tribute.[5] In Britain, this contractual form was considered a self-adjusting system of remuneration that effectively solved the complicated relations "which exist between masters and men."[6]

Mining companies in Cornwall pushed to restrict the tribute system, but skilled workers were able to maintain their work as tributers until the twentieth century in some mining camps with Cornish influence.[7] The same division of labor and negotiation capacity expanded to Southern Australia and the American West, as the Cornish diaspora provided skilled miners to other frontiers in the Anglo expansion.[8] The most skilled miners in the Intermountain West's first rushers were either Mexicans, experienced barreteros who crossed the border back and forth, or Cornish miners, commonly referred to in the literature as "Cousin Jacks."[9] In essence, the tut and tribute and partido traditions largely influenced the persistence of piecework and value-based remunerations in the Intermountain West.[10]

Both Cornish and Mexican systems of payment shared something in common: a corporatist organization of guilds typical of ancien régime societies, and Cornish miners, like the Mexican barreteros, maintained systems of self-organization. Mutual aid societies were common among the Cornish and, as with the tut and tribute, traveled across the Atlantic into the Intermountain West's mining camps. In the second half of the nineteenth century, mutual aid and secret societies emerged in the West's mining camps, primarily focused on treating the wounded miners and supporting the families of the deceased workers in the underground. The largest of these, the Noble and Holy Order of the Knights of Labor (KOL), deeply impacted the construction of a formal union in both the hard rock and coal mines, having successful strikes in different mining camps in the 1880s. In 1890, a section of the organization in Ohio together with the National Progressive Miners Union formed the United Mine Workers of America (UMWA). Former sections of the KOL were vital in the organization of miners from Montana, Idaho, Utah, and South Dakota, who followed the UMWA model and founded the Western Federation of Miners (WFM) in 1893. Among the workers' movement in America, the UMWA and the WFM were the only industrial unions and were confronted in the next years with the American Federation of Labor (AFL). Founded in 1896, the AFL was a federation of craft unions, which implied that different skill sectors into a single firm had different contracts and locals. In opposition, in 1905 the WFM organized the Western Labor Union and founded a new and more radical

organization, the Industrial Workers of the World, with a high participation of socialist organizations.[11]

The transformation of the work process by industrialized corporations in the last decades of the nineteenth century and the beginning
of the twentieth undermined skilled miners' place in the organization of labor. These transformations altered the nature of the political
bodies of workers, but also had an impact into their physical bodies.
As scale economies in the extraction and processing of ores increased,
the selective mining organized by skilled workers like the barreteros,
maestro torteros, and their Cornish counterparts became marginal, and
the tribute system they benefited from disappeared from most mining
camps. Extensive mining produced another result among the miners: a
dramatic increase in the number of accidents underground and on the
surface. Scale economies, especially by using more powerful explosives,
pneumatic drills, and railcar transport of ore, increased the accidents.
Pneumatic drills projected rocks and sometimes broke metal into workers' eyes; stocked rocks and railcars fell and rolled over miners, crushing bones, limbs, and organs.

The care of the miner's body emerged as a central demand along
with the growth of industrial miners' unions in the hemisphere. At least
twenty-five local affiliates of the WFM established hospitals between
1897 and 1918 in the American West, and several companies set up hospitals and doctors in their camps. In Mexico, the barreteros had disappeared in the mining of copper but remained influential in the selective
mining of gold and silver until the introduction of cyanidation technology at the turn of the century. In the first years of the twentieth century, the barreteros in Pachuca and Real del Monte pushed the company
to install a miners' hospital to treat the traumatic wounds of miners'
bodies. The Mexican owners did not complete the hospital project, just
as they could not finish the firm's productive transformation. In 1906,
the US Company's American management inaugurated a new miners'
hospital in alliance with the barreteros. That was the swan song of the
skilled Mexican miners. In a few months they became irrelevant in the
firm's work process.[12]

This chapter argues that this transformation, from mutual aid societies to radical industrial unions, was not the product of a few radicalized and ideologized leaders but an answer to the changing nature of

the mining business.[13] Moreover, the development of these autonomous political bodies was deeply connected with the changing division of labor and the effects on the worker's body. The metaphor of Jennings took some elements from a long-standing tradition, but it also revealed the changes in the workforce and its relationship with the form of the corporate Leviathan. In short, the transformation of the technology of mining inside big corporations, generating economies of scale in mining and processing, required a radical organization change in the mining operations. These mutations in the labor conditions of industrial mining formed a different labor force than the one necessary by traditional mining, eliminating some functions and establishing others inside the work process. The corporation mutated the body of workers internationally, both as a political body and as a physical one. This chapter will analyze the different transformations that this body of workers experienced during the decades after the multinational corporation's emergence.

CHROME, NEW JERSEY

Around the first week of April 1918, thirty-three workers of the US Company at Chrome walked out of the US Metals Refining Company plant in protest against the firm's unwillingness to negotiate a 6.25 percent increase in wages. Although they were a small minority of the firm's 1,200 workers, they were firemen and furnace workers, who were among the most skilled and strategically important jobs in the electrolytic refinery. In a few days, and despite the firm's attempts to keep the operation running, the corporation had to shut down the plant.[14]

The wildcat strike ended the firm's capacity to maintain operations despite the labor unrest of the first decades of the century. On one side, the corporation had internalized a considerable part of the world market by establishing its commercializing networks and input companies, but the productive integration dislocated some essential activities. The processing plant was a centralized unit with significant economies of scale, unlike the small and medium-sized plants in the mining camps. Typically, refining plants were in a different location than the processing plants, especially in industrial metals. In the case of the US Company, the copper, lead, and silver were extracted in the mines in the mountains and canyons, in Mammoth, Bingham, and Real del Monte, then processed in blast furnaces in Midvale and Shasta County and the cyanidation plants in Real del Monte. The Mexican cyanidation plants and the smelter in Shasta County, California, also had electrolytic refineries for producing bullion, but the majority of the concentrates were shipped to two different refineries, the Chrome copper refinery in New Jersey and the Grasselli lead refinery in northwest Indiana. These two last divisions concentrated a big part of the firm's concentrates and some additional metals from smaller firms nearby. The final products of these last plants were then sold by the company's selling agency in New York, along with the Mexican division's silver and gold bullion that traveled by railroad to the north. In this corporate division of labor, each division established new mechanisms for the coordination of processes, changing the workers on a local level. In short, the international corporate body required the coordination of bodies of workers transnationally.

There were three primary spaces of differential operation of the firm. The most traditional setting, with the most ingrained identities, was found in the mine. The new technologies of extraction redesigned the metabolic process between the ore bodies in the underground and the surface's processing plants. More ore was extracted from the mine, and more energy was required inside the mountain, increasing the use of explosives, pneumatic drillers, electricity for hauling and lighting, and drainage systems. The consolidated mines, a product of the expansion of the corporations, connected with other properties in the underground giving birth to underground cities. On the surface, more matter was processed with increased amounts of energy and the use of a rationalized series of processes connected by conveyor belts. Crushing plants reduced the ore into small particles, concentration plants eliminated the components with too little metallic value, and cyanidation plants and blast furnaces transformed the sulfides into metallic oxides. Finally, refining plants transformed the metallic concentrates, on solution or dust, into metallic bullions. To be sure, the amalgamation with mercury and the processing in small furnaces required differentiating between high-grade ores and low-grade ores, which was therefore a specific skill of the underground workers in order to maintain the profitability of the operation. The cyanidation plants and the large blast furnaces used by mining corporations (as the US Company used in the beginning of the twentieth century) eliminated that need and favored, in contrast, the massive exploitation of low-grade metallic ores through the use of big infrastructural networks that connected the mine with processing and refining plants.

This new industrial organization implied an increased relevance of certain types of skilled workers. From the bottom up, in very literal ways, the new system required less-skilled workers who executed the infrastructure and the extraction of the rocks, and the new ultra-skilled workers, electricians, and mechanics, who supervised the bottlenecks of the metabolic process with the mine. Moreover, while the selections of the ore by the underground workers became largely irrelevant, some of the workers at processing and especially refinery plants gained extraordinary importance. Their smaller number and more centralized geographical location allowed them to integrate into compact units. The relatively less skilled labor at the mining camps was more easily replaceable, and work stoppages were not costly for the firm. But at the end of the decade of 1910, the firm faced strikes in more central and critical spaces of the production Leviathan: the refineries. The first serious and generalized labor problems that the corporation faced hit precisely in the final steps of the firm's productive system where skilled labor had taken a new, central role. A wave of strikes affected the multinational metallic Leviathans between 1918 and 1923, stopping the operation in strategic places of the firm's industrial structure in several spaces, from East to West and North to South.

The wildcat strike in New Jersey, in consequence, had structural origins, but in the previous years, labor activists had been able to organize a select group of skilled laborers. Since 1916, the chemical and metallic firms in New Jersey had increasing labor troubles, and some strikes had erupted among the machinists and firemen trades demanding eight-hour days. In January 1917, a wildcat strike took place in the ASARCO smelter at Perth Amboy. The workers refused to take the advice of the AFL organizers who arrived to mediate a negotiation.[15] By the end of 1917, the chief of police of the industrial area, Chief Perry, tracked some "western agitators" (allegedly Wobblies) and tried to isolate them from the working men. He publicly said that he was eager to follow all three "western methods" to deal with the organizers: "First, a warning; second, an applied coat of tar and feathers, and free transportation across the state line; and third, lynching."[16]

The agitation concerned the US Metals Refining Company's managers, subsidiary of the US Company, and at the beginning of 1918 they increased, preemptively, the workers' wages. The new division of labor within the corporation and inside the plant had given specific workers a greater capacity for halting the plant's operation. The furnaces had to remain active to continue production, but the stoppage generated a high cost of maintenance and restarting of operations. The centrality of the work of the men provoked the stoppage of the plant in the next days. As the company was an important supplier of copper for the war effort, the Department of Labor sent the former US representative from Pennsylvania, J. J. Casey, as a mediator, to negotiate an agreement between the firm and the striking skilled workers. Strikers also walked out of the installations of two other refineries seven miles to the south in Perth Amboy: the American Smelting Company and the Raritan Copper Works. On April 13, 1918, the striking workers took two of the government officers to Newark, which was probably the organizing center of the region's conflicts, in what the press called a kidnapping. After the incident, the negotiations sped up, and in the next week, the workers and management of the US Company reached an agreement.[17]

It was a significant defeat in the firm's strong anti-union policy and the company tried in the following months to reaffirm its sovereign capacities. Despite the labor agitation, the division in New Jersey had increased its value more than the surrounding factories and smelters of the area. It used that new power to recognize and assert its sovereignty over the workers' body.[18] In July 1918, through an official communication to the city mayor, the United States Metals requested the appointment of twenty-seven of its men as special marshals. The company gave the names of the department heads in the plant's different sections. This increased capacity of the firm to impose direct violent rule over the body of workers required obtaining legitimate power of exercising violence from the state. The request of the firm was supported by the Military and Naval Protection Bureau, as the division was a significant supplier of copper for the manufacture of weapons to the factories in the vicinity.[19]

At the beginning of 1919, the corporation faced exactly the same problem on the West Coast, the Shasta Smelter of the Mammoth Copper corporation in California. Once one of the greatest copper producers in the West, the mining area had had declining returns in the last years, and the smaller companies had closed. Only the smelter of the US Company's division remained in business and treated the ore of the Balaklala and Keystone areas as well as the company's own ore. The key metal extracted from the ores was copper, but the electrolytic refinery at Shasta had allowed processing of other minerals from Northern California, like cadmium, zinc, and gold, in a plant of six hundred workers. On May 7, 1919, thirty boilermakers, electricians, machinists, and railwaymen presented demands to the division chief, G. W. Metcalfe, for an increase in wages and to reduce their Sunday hours to a half day. These workers, a minority of all employees, as with the strikes the year before in New Jersey, knew about the firm's inability to maintain operations without them.[20] However, management refused once more to negotiate with them. This time, the company's engineers had new leverage: the falling price of copper. Metcalfe declared that the company had operated at a loss for the last five months and threatened to lay off all workers of the firm if the workers walked out.[21]

On May 11, the thirty-six mechanics walked out of the plant, and the firm continued its operations, with little hope of being able to maintain them for long: only two days after the walkout, the company had to close one of the three blast furnaces in the factory. While the mechanics and boilermakers did not directly stop the furnaces and machines, their normal work was to fix the usual stoppages and broken machines in operation. On the evening of May 14, the strikers convinced half of the locomotive crane operators not to come back to work the next day. Metcalfe, even though he had assured the day before that he would not hire strikebreakers, brought in some replacements for the striking mechanics. However, without the crane operators, the smelter would run out of raw mineral soon. The other big player in the area, the Balaklala Copper Company, preemptively raised the miners' wages that day, but their only option for smelting their ore was in the US Company plant. On the morning of May 17, only six days after the walkout of the thirty-six mechanics and boilermakers, the Mammoth Copper Smelter shut its doors, and all the mining firms in the district quickly followed.[22]

In the following weeks, hundreds of miners left the district, and the stoppage extended to some gold operations in Nevada, which also shipped their concentrates to the firm. The situation did not change aftet that. Copper prices remained very low, and the US Company left only twenty workers to do some development work in their properties and in the idle mining territories. By the end of the summer, the company only kept several watchmen, and two managers and the company's doctor were relocated into the corporation's operation. In those months, the multinational company objected to its tax valuation and tried to get reduced taxes on all the divisions tied to the production of copper, from Shasta County to New Jersey.[23] On the eastern side of the continent, the firm tried to repeat the new strategy against the critical workers.

On August 19, 1919, the skilled workers at the plant of the US Metals Company at Chrome again went on strike. This time, as with some other postwar strikes, the owners were able to resist longer. The strike lasted over a month, but in the end the workers again achieved a settlement with the firm.[24] As a preventive measure, the firm sold some shares of the US Metals Company, losing its position as a majority investor, but the administration of the company stayed among the corporate holding. Other links in the production chain, from the end product to the raw materials, were struck with more success in the next years.

MAESTRANZA, PACHUCA

The installation of the Guerrero and Loreto cyanidation plants at Real del Monte and Pachuca completed the company's movement toward extensive mining in those locations. As a result, the skilled Mexican miners, just as the Cornish in North America, were marginalized from their control in the production process and from the new organizational form. The body of workers required for the new exploitation was reformulated in the next years by the US Company's management. With the arrival of the US Company, tribute and partido disappeared in the locality, but not the tutwork. The system of relatively independent miners at the margins of the firm's exploitation remained in the Mexican division on the *Terreros* properties. The Terreros were deposits of waste, either processed mineral or discarded rocks on the company's property and still owned by the corporation. Under a concession, independent miners explored these deposits and evaluated the rocks and crushed ore with enough metallic content to be processed with the cyanidation process. Another rate system at the operation was more direct. The firm employed crews of miners, most of them old workers of the corporation and miners doing some extra hours, to extract cars of waste and to clean or build infrastructure in the tunnels. These contractors, or *destajeros* in Mexico, remained under the firm's direct employment and were controlled by management, but they kept some organization features of the old tutwork systems. They had a leader of the crew who made the labor arrangements with the firm, although they worked for flat rates and, at least in the first years, had little to no power of negotiation, as their work was less skilled.[25]

The skilled workers were concentrated, as in Mammoth's exploitation, in the workers at the cyanidation plant, as well as the mechanics, boilermakers, electricians, and railway workers. They had a different relationship with the industrial growth of the country in the last decades, as most of them had worked in other industrial firms earlier and came from urban communities. In Mexico, most of the rural population came from the indigenous towns around the area in Hidalgo, and from a community of farmers from the indigenous towns in the western state of Michoacan. In contrast, the urban workers came from central Mexico, which was more industrialized, following the railroads that connected central Mexico to the Veracruz port to the east and the American railroad system to the north.[26]

During those years the firm struggled not only with the revolutionary war, and the shortages of explosives due to the war in Europe, but also with the adjustment of the industrial structure of Pachuca and Real del Monte to the demands of the new, multinational corporation. In the first decades of its operation in Mexico, the firm extended the compressed air system in the mines and modernized the crushing and grinding machines, requiring increasing volumes of power. The crushing plant at Loreto operated twenty-two hours a day, processing around two hundred tons of ore per hour. At this level, overloaded machines caused a loss of 6.6 percent of the working day through repairs and replacement of parts. The cost of the spare parts, and the transportation required for their importation, forced the company to produce and repair more of the pieces than in any other division. The mechanics producing these extra parts had a workshop, La Maestranza, in Pachuca. It was located between the railroad station and the Loreto mill, near the firm's warehouse. The Maestranza concentrated the modeling department, foundry, mechanic workshop, and the boilermaking trades, sending the workers and the repaired pieces as needed to the several stations of the firm, from the underground to the small electrolytic refinery. In this strategically critical division of the firm's operations, as in the conflicts in New Jersey and California, the company had the first important strike in 1923.[27]

Agitation among the mechanics in Mexico started around the same years as the strikes in Mammoth and Chrome. In 1919, the Casa del Obrero Mundial members at Pachuca, who had protested in 1916 for payment in gold, had asked for a 50 percent increase in the wages of the Maestranza. Covacevich, chief in the workshops, and the firm's management dialoged with the small number of workers and rejected the petition without further labor conflict. Two years later, a different miners' organization asked for the same demands. In August and September 1922, the Unión de Mecánicos Mexicanos (UMM) representatives started negotiations with the subsidiary of the US Company in Mexico for the establishment of workshop regulations at the Maestranza in Pachuca. The UMM was a mutual aid society formed in 1900 in the workshops of the National Mexican Railway. The union had grown in the years before the 1907 Cananea strike, but the heavy anti-union policy after the miners' strike pushed the organizers into the shadows, and they moved their organization to Chihuahua. In 1912, the union had organized a successful strike in the railway company, leveraging the workers' mobilizations during the revolution, involving around twenty thousand railway workers from Chihuahua to Aguascalientes, and demanding a *Reglamento,* a national set of workplace rules negotiated between management and workers. After the success of that strike, the railroads' integration of mines, and the usual geographical proximity of the mining workshops to the railway stations, the UMM organized mechanics in the mining operations. The Pachuca workers were the first experiment in leading a collective negotiation with the extractive multinational corporations.[28]

Management of the division agreed to negotiate with the union leaders, and the talks continued during the next three months. By the end of November 1922, the union threatened to go on strike with the Welders Union (Unión de Moldeadores), and the company headquarters in Boston became involved. The technicians in Massachusetts had experienced the national wildcat strike of switchmen in the spring of 1920 in the United States, which had affected, although briefly, the shipment of ores from their divisions in East Chicago and Midvale.[29] These actions forced the multinational Leviathan to acknowledge, at least implicitly, the growing strength of the mechanics and boilermakers, but the firm was unwilling to accept the growing strength of union leaders in the Maestranza. The company relied on the government and the Local Arbitration Board's support, while the union had the support of a pro-worker discourse at the federal level.[30]

The major dispute was about three of the more than one hundred regulations in the workshop: a general rise of wages, seniority rights, and union recognition. Soon, the negotiations, which included Mexico's industry secretary's local and federal agents, broke down between the firm and the union. The corporation fired seventeen employees of the workshops who had participated in the locals, and in the last week of November, the UMM made a strike vote as a reassuring measure against the firm. The tension grew with the local authorities. Soon, the governor rejected acting as an arbitrator as the federal authorities asked the firm's management to negotiate in Mexico City. The conflict regarding the sovereignty of the negotiation escalated in the next days, as the managers ignored the labor department's requests to appear at their offices. On January 4, 1923, the firm published the *Reglamento para los talleres de Maestranza,* ignoring both the union and the federal government. Two days later, the UMM representatives sent a telegram to the state governor, local authorities, and the federal government announcing the walkout from the Maestranza during the day shift.[31]

As with other mechanics walkouts, the workers' capacity to stop the production flow between the production units resulted in questioning the authority of the managers over the work process. The mechanics walked out of the Maestranza and the mills at Guerrero and Loreto, blocking the doors at both plants and the general warehouse. On the second day, the workers convinced some blacksmiths to walk out of the mining operations in Real del Monte. The firm could respond by pushing laborers into the crushing and cyanidation plants' works, but without enough supplies or mechanics to fix normal problems that arose. By the third day, the small crew of workers who could reach the Loreto mill had finished the unprocessed ore reserves and had to stop operations. Workers stopped the general manager and his assistant, D. S. Calland and M. H. Kuryla, from accessing the Loreto plant until they could search for strikebreakers. The managers' ultimate humiliation was that, after reclaiming their power as managers, the workers proclaimed that "they do not recognize any other authority but the Strike Committee." Jenson Smith, the superintendent, drove over one worker who tried to stop him from getting inside the plant.[32]

On the fourth day, strikers impeded the company's trucks from leaving the general warehouse with forty-five silver lingots. Two members of the strike committee and two mounted police officers stopped the delivery and forced it to return to the Loreto mill. Only after negotiations with the governor could the company transport the bars to the railway depot. A week later, the vice secretary of the interior, Gilberto Valenzuela, and the labor department of the secretary of industry arrived in Pachuca from Mexico City. Despite the intervention of Secretary of the Exterior Alberto Pani on behalf of the company, by January 15, company officials were forced to negotiate in Mexico City with officials from the office of the interior and strongest contender to the presidency, Plutarco Elías Calles. The strikers demanded a 50 percent wage increase, the enactment of seniority analogous to the national railways, union recognition, and a total payment of expenses and wages for strikers. After the negotiations, the union officials only obtained payment of expenses and wages and union recognition, and Calles demanded the reestablishment of seventeen dismissed workers.[33]

The negotiation itself tasted like a complete victory for the UMM. Only twelve days after the start of the strike, they had won union recognition and gotten their expenses paid, and the company had restarted operations. The Hidalgo strike was a model for the ensuing organization of mechanics by the UMM in other mining multinationals. In the next years, the union negotiated a similar contract with the Mexican Zinc corporation, a division of ASARCO in Coahuila. As a response to the strike, the United States Company's management deepened unionized workers' centralization in the Maestranza, and they increased the reserve of supplies underground. This implied restricting the mechanics' strength to the workshops, farther away from the plants, and that they would be better prepared for a lack of supplies in the tunnels. They developed a better road and rail network of the mines underground in order to reduce the territorial control of the mechanics.[34] Nevertheless, neither the structure of the conflict nor its actors were unique to the mining district in central Mexico: the divisions in the continent kept facing conflicts from the same bodies in other latitudes.

HIGH FURNACES, MIDVALE

"We do not wonder that anybody would rather go anywhere than stay in Bingham, which is the most repulsive mining camp that we know of in the U.S. We do not deprecate its unfortunate inhabitants, but refer rather to its physical conditions. . . Bingham has been most fittingly described as 'A sewer four miles long.'"[35] The editors of the *Engineering and Mining Journal* referred in this manner to the largest copper deposit in exploitation in 1912. The reshaping of the large corporations' workforces in the area had provoked great agglomeration of settlement in the canyon around Bingham Creek, used by the inhabitants as an open sewer. The observers believed that the disease and dead among the miners' community were not higher only because the corporation also expelled toxic sulfides to the creek, which disinfected the human waste in it.

As in Pachuca, most of the area workers worked for multinational corporations. The US Company allowed independent miners to explore the firm's old properties and, at a flat rate, keep the concentrated metallic ore that they could extract. This permitted the firm to directly explore the claims with a high profitability rate and maintain indirect exploitation of the less productive ore. Under the system, management kept control over the inputs, instruments, technical assistance, transport, and processing of the ore. Most of the *independent* miners were workers of the firm searching for extra pay or old-time miners who could not find direct employment in the corporation. They were a marginalized yet essential body of workers in the corporation, vital in the firm's new division of labor.

This marginality of the independent miners and the unskilled nature of the work in the underground often translated as racial segregation. In Mexico and the American West, mining corporations segregated the work and workers' benefits by race, using fragmentation of the processes to prevent class solidarity among miners. In Utah and California, the *padrones* system worked to control immigration to the West for the firms' benefit. The padrones used ethnic and immigration networks to control new arrivals' lives, took part of their salary, and provided them with lodging and credit at the company store, restricting their capacity to rebel against working conditions. The padrones positioned these unskilled workers against white Anglos, who could better access the strategic jobs at the firm. In general, wage and job discrimination were standard in the metallic companies in the West of the United States and the north Pacific region of Mexico.[36]

Immigration, a product of these practices, built a distinctive architecture for the town. By 1912, around 65 percent of the residents were foreigners, who had arrived in several waves. Anglo and Irish communities occupied downtown; Finn town was closer to the Carr Fork mine and was mostly populated by Finns, Norwegians, some Swedish, and Italians. The Highland Boy Cliff was named Little Austria, replicating the complicated ethnic geography of Slovenians, Croatians, and Serbians. To the north, around 1,200 inhabitants formed Greektown, plus a small community of Japanese.[37] Although this complex city had internal tensions, most of the conflict existed as the rivalry between the mountains' migrant miners and the farmers of the valley who were members of the Church of Jesus Christ of Latter-day Saints (LDS). They represented a different relationship with space, both symbolically and materially, and corporations would employ Mormon farmers seasonally and as an industrial reserve army in moments of labor conflict. The rivalry between the two communities, which involved stereotypes about Mormons' polygamy and the alcohol abuse of gentiles, fostered the economic and symbolic disputes over space and erupted periodically as racial violence episodes.[38]

The challenges of organizing the large numbers of miners underground determined the nature of the workers' protests in the second decade of the twentieth century. In the fall of 1912, the WFM made a strike vote, 9 to 1, for a walkout, asking for an increase of fifty cents a day, a 17 percent increase, for over three thousand workers in the different firms in Bingham Canyon. The control that padrones had over the life of Greeks in the mining camp fueled a substantial part of the workers' animosity against the firm. The conflict followed the same pattern as others in the metallic West, both north and south of the border. Along with the Associated Union of Steam Shovel Men's secretary, the president of the WFM started to organize solidarity strikes in Nevada and New Mexico in other companies of Daniel C. Jackling, the engineer who was a general manager of the Utah Copper Company. On its side, the companies made a unilateral increase in wages of twenty-five cents and refused any negotiation with the union. As in the other conflicts, Governor William Spry sent 350 men to the area, paid by the corporations, to contain the strikers and stop them from blocking access to the mines. One week after the beginning of the conflict, the corporation, the civil authorities, and the churches, both Mormon and Greek, paved the way to end the strike. The governor and a bishop of the Greek Orthodox Church called on the workers to stop the blockage of the firm's activities, and the National Guard inspected the camps to intervene in them. The firms fired the padrone Leon Skliris, who had provided the workforce in south Utah after the strike in 1903 and controlled the Greek communities across the state. The gradual conquest of the space, in both military and ethnic terms, and the unilateral increase in wages and firing of Skliris, fostered the capacity of the firms to restart operations in the next days.[39] A similar defeat and the union's rivalry with the Industrial Workers of the World (IWW) provoked the destruction of the WFM in 1914. That year, after a physical confrontation with the leadership of the Wobblies, which ended in the destruction of the union's offices, both organizations were expelled from Butte, Montana. In 1916 the WFM changed its name to the International Union of Mine, Mill, and Smelter Workers (IUMMSW). It later reintegrated into the American Federation of Labor (AFL), becoming, along with the United Mine Workers Association (UMWA), one of the few industrial unions in the federation. At its founding convention, held in Denver, the United States mine miners of the US Company in Bingham formed the local 2.[40]

In spring 1923 the United States Company, along with the Utah Copper, ASARCO, and Anaconda, faced Wobblies' agitation in the metallic mines in Utah. In May 1923, the IWW organized a massive walkout of Bingham miners, coordinating it with several other western strikes. Although most companies could operate on a reduced scale the day after the event, they had to rely upon less- and non-skilled miners, which diminished their productivity. Throughout the year, management of the Big Three (composed of Anaconda, Utah Copper, and the US Company) tried to rebuild productive capacity, attracting workers by improving living conditions in the canyon, building boarding and change houses, and making a preventive fifty-cent increase in wages. Managers tried to purge the radical elements and started conversations with Bingham's mayor to prepare the sheriff's forces for a new strike in May.[41]

After their success in opposing the strike in Bingham, the companies had planned to reduce smelters' wages in August, which invigorated agitation among the more specialized workers. Like in Pachuca, the relationship between the mines and the mill caused some workers to be in a stronger position to stop the company's productive operations. The relative smallness of the mill compared to the mines, the scarcer labor for manufacturing the ores, and the company-specific work processes allowed a stronger organization at the plant. These new actors—the industrial system's crafters—could organize a complete stop in production, which underground workers could not.

At the end of March, the IWW committees demanded a general wage increase in the mills of Garfield (from the Utah Copper Company), Murray (from ASARCO), and Midvale (from the US Company). After the companies rejected the demanded increase of fifty cents, Midvale workers walked out of the mill on April 12, 1924. Unlike the UMM, the IWW maintained their no-contract policy, as they considered contracts a threat to their right to strike. However, as in Pachuca, the strategy was to break the internal coordination between the corporation's productive units. The work stoppage immediately interrupted the chain of ore processing, precisely at the point of transporting it from the mines to the smelter where lead and copper were separated. The furnaces stoppage was expensive for the company, as they were hard to reheat, and the melting chambers would become congested with cold ore. Once again, the high costs of the shutdown, the difficulty in replacing the entire plant workforce, and the work stoppage that was also happening in the mines sped up the end of the conflict. On April 16, the workers had won a twenty-five-cent wage increase and were back to work only four days later.[42]

If the economies of scale pushed management to further integrate the different stages of extraction and processing of the ore, the emergent power of strategic workers pushed them in the opposite direction to enhance the company's productive units' autonomy. These two forces were behind the growing anxiety of corporate officials as Jennings, of the sovereign power of the firm over the body of workers. As strategy, company officials increased supplies at the mines, instead of the general warehouse and ore in the plant. Furthermore, they accelerated the work on some improvements in the living conditions in the Bingham district. Those measures were effective in fighting a subsequent strike at Bingham in May, although, as happened in 1923, the corporations faced costs in replacing the workers during the following months. Despite their failure, IWW organizers in both mines and the smelter pushed the companies into making another preventive increase of fifty cents per day by the end of the year.[43]

This last strike was simultaneous to the negotiation in Hidalgo and shared the structure of the walkouts in California and Indiana: they attacked the centralization of production and productive units' interde-pendence. The workers' strategy was to cut circulation at the different stages and spaces of production; the interruption of the labor of a single group of workers could achieve this. Despite the differences in government intervention policies, strike contention during World War I in the United States, and the promotion of workers' organization in Mexico during the revolution, the workers experienced and comprehended the respective industrial relations systems in very similar terms. Before and during the conflict, only the corporations' management acted and reacted in a coordinated fashion, but local action of the strikers formed a non-explicit and most likely unconscious coordination. In the coming years, companies altered their productive strategies to combat mechanics' and smelter workers' new strategic position in labor conflicts.[44] Simultaneously, workers' unions established a new system of conscious coordination that mirrored the organization of management in mining companies in Mexico and the United States.

The wave of strikes in the different divisions of the US Company between 1918 and 1923 revealed an increasingly conscious body of workers in the local scene and a dispute over the internal relations inside the corporate Leviathan. It showed that the construction of the body of workers was not unilateral, from the management to the laborers, but a contested negotiation of roles and rules. Moreover, the territorial dispute inside the working place challenged the dominium and emporium of the engineers and managers over the production process of the firm and the political nature of the division of labor inside the firm. However, as a result of this conflict, there was an expansion of corporate sovereignty over the space of reproduction of the workforce: the living quarters of laborers.

CHATANIKA, ALASKA

In April 1930, the *Fairbanks Daily News-Miner* reported some prospectors' incursions into the Beaver District, to the north of the Chatanika River, looking for gold in the territory. They were old-timers, prospectors who had arrived in the rush twenty or thirty years before, and had stayed in the territory looking for their fortunes. In the first decade of the century, Alaska was considered an extension of the American frontier and the destination of miners searching for their destiny, the space of realization of personal ambitions and independent life. Rushers still operated in Alaska until the third decade of the century, following their gut and their projections of finding the yellow metal. This time they predicted mineralization in the area between the Tanana and Yukon Rivers and had to use dogsleds because the snow blocked the small roads to the thawed area. Some other prospectors were pushing the boundaries of mining to Kobuk, between Tanana and the city of Nome, which was only communicated through trading posts. Most of the local population was indigenous.[45]

In the first decades of the twentieth century, the workers established not only an autonomous operation but also quasi self-regulated societies. Miners were in charge of the definition of property limits, government, and the exercise of legitimate violence, through courts or physical retributions.[46] As they expanded the frontier's limits, the mining communities were self-limiting and self-regulated. However, independence and autonomy were a distant memory by the 1920s. By 1930, even the most stubborn old-timers prospected with the only aspiration of finding a profitable dredging camp that could be sold, afterward, to the Fairbanks Exploration Company. The small crews panning and making small pits did not expect to hit it big, but only to find an extra source of income of their employment in the activities around Fairbanks, which revolved around the US Company's divisions. The explorers following the Chatanika River to the Pacific were sometimes accompanied by engineers of the US Company's subsidiary, as they worked indirectly for the firm.

When the company arrived, independent miners became subordinated to the capacities of dredge technology. They combined the exploration of rivers during the spring's first days with seasonal work during the summer in the mining camps. In April, the Fairbanks Exploration Company at Gilmore and Goldstream's camps had already resumed operations, along with the foundry, power plant, and offices at Fairbanks. However, the mining camp at Chatanika remained inaccessible. In the second week of the month, the company and the Alaska Road Commission jointly sent tractors and plows to break the road over the hills to the Chatanika mining camp, the farthest north dredging operation of the firm.[47]

During winter, a single watchman guarded the camp. He was isolated for a minimum of five months and took care of guarding the property that had to be abandoned due to the extreme weather. Smaller camps were usually seasonal, and workers stayed in tents during the firm's months of operation. The extension of the dredging activities around Chatanika pushed the firm to install well-equipped installations for the miners there in 1925. Two galleries of board houses were connected, through an aisle, to a collective diner. There was a water heater and a change house on the barracks side and, to the south of the camp, were the latrines. The chief of operations lived in a fully equipped house to the west of the dining hall, which the watchman used during the winter. The production buildings were mixed with living spaces. The workshop was to the east of the boardinghouse, the assayer's room was by the chief of operations house, and the storage was located to the side of the change house. Around fifty yards from the camp, upon the hill, was another living building: a school for the workers' children.[48] As alcohol was prohibited in the camp, the firm assured that the miners could access bootleggers and illegal salons in the camps' outskirts.[49] In short, the firm defined the limits of mining in the area and the borders of the legal and illegal: corporate sovereignty over the territory was, basically, completely unchallenged.

The transformation of sovereignty in the workplace to sovereignty in living spaces was not something new for the firm, nor was it restricted to Alaska's frontier territory. On the one hand, it was deeply inspired by a tradition of corporate power and company welfare in mining communities. On the other hand, it was a growing concern in controlling the formation and rules of mining communities integrated in the international division of labor implemented by the firm. The firm's control over the urban spaces where the workers lived was often reflected by the combination of corporate sovereignty with state power.

In 1919, the Mexican division of the US Company owned seven

buildings that were used for government agencies in Pachuca and Real del Monte, including the schools at Omitlán and Guerrero, two tax offices, the road administration in Real del Monte, the prison, the civil Pachuca hospital, and the church in Real del Monte. In the years before, the firm contributed to the operation of some of these buildings, especially the civil hospital, and charged symbolic rents for their use. The other forty-five nonproductive urban properties of the firm reproduced the corporate body's social structure in the workplace. While the least-skilled workers lived, usually, in improvised camps in Real del Monte's outskirts, the skilled workers could live in the company's properties. The clerks and the aides in charge of paying wages and those in charge of the stores usually lived in a small house by their workplace, the storage at Maestranza, and the office at Las Cajas, both in Pachuca at a minimum rent. The company owned twenty-five *vecindades,* where several families could live in small apartments sharing different spaces in the different working places: Pachuca, Real del Monte, Guerrero, Velasco, and Omitlán. The firm collected rent from the corporate payroll properties and established rules for the workers who inhabited the spaces. Gambling, prostitution, and alcohol were forbidden. This corporate welfare for the technical workers increased in the next years, as an answer to the labor agitation of the beginning of the decade of the 1920s. By the end of the decade, the corporation stopped collecting rent from the rooms given to their workers, extending the benefits to carpenters, masons, rail workers, and their aides. It was an answer to the organization of workers, and it tried to prevent further radicalization among skilled laborers. In these living spaces, the corporation tried to control the rules of socialization and forbade any political reunion of any sort in any of the firm's vecindades.[50]

Regulations over the mining communities' lives, the installation of corporate houses, and other corporate welfare measures became a corporation policy in its different divisions, although adapted to the local conditions. The US Fuel Company, the division of the US Company in southern Utah, claimed to have the best mining towns in the area, in East and West Hiawatha. The firm had previously installed around one hundred houses, boardinghouses and change houses, a hospital, and a school. The extension of the firm's sovereignty into the town was similar, in this way, to the functions of the company in Pachuca, but with an even greater political control over the administrative authorities. The town outside their Panther mine was incorporated as "Heiner" in honor of the superintendent, Moroni Heiner. The firm was involved in the spiritual aspirations of the workers. They had financed the first dominical Mormon school, and in 1924 started the construction of a church, provided with electricity, drainage, and running warm water. The corporation contributed to the elementary school and the junior high school in Hiawatha, paying for the teachers, provided transportation to the high School in Price, and coordinated the local parent-teacher association. In Hiawatha the firm installed saloons, bowling alleys, and dance saloons, and organized projections of movies. In sum, the corporation tried to both provide and control all the workers' needs at the company coal towns, from the spiritual to the material, controlling the different spaces of living of the subjects.[51]

To be sure, the construction of the urban space in mining communities responded to the larger endeavor of the corporation building a body of subjects in their different divisions. In Bingham, the company had established partnerships with padrones and actively promoted workers' ethnic segregation,[52] but the increasing labor unrest in the 1920s pushed for a new kind of relationship with the living spaces in the canyon. Just as the conquest of the space underground pushed a consolidation of mining claims, the corporate welfare endeavors of the firm in Bingham appeared as a rationalization of the initial anarchy of the mining community in the canyon. The establishment of improvised camps and the growing number of workers in the area had produced several urban problems. In 1920, Bingham counted fifteen thousand inhabitants, and the ethnically based architecture of the town was severely overcrowded. The congestion of houses and activities had provoked two enormous fires in 1923 and 1924, and a landslide, which had cost almost half a million dollars and thirty-nine human lives.[53]

The US Company was a pioneer among the firms in the Oquirrh range that started to build living complexes for their workers. In the summer of 1923, soon after the first walkout in the area, the firm started the construction of corporate change and boardinghouses outside the Lark mine, one of the two large mining pits that remained after the consolidation of mining properties, and that was communicated by the Yosemite tunnel to the Bingham town. The contrast between the corporate plans and the urban arrangement in the mining communities was remarkable. The houses and businesses at Bingham followed the creek path, and the accidents around the mountain implied an irregular distribution of population in several stops, elevations, and sinuous roads. In contrast, the barracks and changing houses were arranged in squares, on a flattened surface, following the company property line, and with straight roads that allowed the transportation of materials. The US Company installed drainage, running water, and heaters in the camp. In the following years, it installed other boardinghouses and cottages of four rooms, integrated restrooms, and expanded the town's urban services. By 1930, Lark town had a company hotel, a Mormon temple, a ballroom, and a saloon. The main street had several stores, a school, a tennis court, and several houses reflecting the workplace hierarchies. The boardinghouse with eighteen different barracks was built to lodge single unskilled men, while the skilled workers could live in twelve cottages with gardens, and even in duplex houses. By the beginning of the 1940s, the boardinghouses were replaced by an expansion of the existing cottages and apartments, and the firm erected a community building for meetings and another recreational dance hall in front of the school. By then, the company town was the only real option for lodging in Bingham, replacing the old Bingham town entirely. As a mirror of increasing control over workers' living spaces, the mining companies absorbed the workers' spaces in literal terms. The ores in Bingham Creek became profitable to exploit with the installation of flotation technologies in the valley's smelting plants, and the town was slowly but unstoppably engulfed by the open-pit mines. The material and organizational Leviathan had swallowed the workers' space, in this even-stronger definition of its sovereign limits.[54] In the 1930s, the firm also increased its interest in controlling and surveying the physical body of workers through medical technology.

LABOR DEPARTMENT, LORETO MILL

On October 17, 1933, Felipe Garcia Contreras walked into the Loreto mill in Pachuca, looking for the labor department office. The labor departments within the firm's subsidiaries had grown over the years, as a counterpart to the firm's increasing control and interest in regulating the hiring processes. The foremen used recruits irregularly, as the multinational corporation doubled down on constructing a homogeneous force of workers in all the divisions. Felipe was nineteen years old, single, and the only means of support for his mother. There is no information on his father's fate, but the clerk registered the family as native to the area and living in the neighboring town of Mineral del Chico. At twelve years old, he started working as a *morrongo,* a child helper in the Tiro Alto mine, but after the crisis in 1930, he had to abandon the mine and labored as a construction worker. The previous year, he was employed in the San Rafael Company, one of the last competitors of the US Company in the area. This time, he applied for the lowest-paying job in the division: truck loader. He appointed his mother as the potential beneficiary in case of death. The clerk at the labor office collected all this information under a standard form, took his fingerprint, and glued a photo of him on the application's left side. He looked very young in the picture and opened his eyes too much. Engraved on the top of his face is his applicant number: 17386T.[55]

By then, data cards of workers were a common practice for all the firm's divisions. In 1925, the labor department at Hiawatha designed the first draft of a written working application, establishing nationality, citizenship, dependents, previous experience, unionization, and health status, as well as requiring applicants to sign the mine rules. These forms were an institutionalization of the controls over ethnicity, union affiliation, and physical characteristics implemented as hiring techniques in the past. For years, the metallic division in Utah had computed the quantities and turnover of rustlers, the new arrivals, and the unemployed gathered regularly at the mines' entrance looking for some job hours. The company officials maintained a weekly register of their number and included, in the biweekly reports, observations of this floating workforce's ethnic composition and union affiliations. As a policy, the firm never employed all the rustlers at its gate, to portray an image of labor's overabundance, even if that entailed looking for farmers in the valley during the winter. By the end of the decade, the registration of these workers had become much more rigorous. The system of individual scorecards, with fingerprints and photographs, was implemented by 1933 in all the firm's divisions. Under the new centralized system, the labor departments in the productive units filed all the scorecards, from the seniors and the discarded workers, in a centralized archive by division. Before any hiring, the labor department had to check previous records of the applicant in their archive to vet any potential problem with the worker. This individualization of the relationship between the multinational corporation and the individual worker allowed the company to centralize and rationalize the mechanism of promotion, punishment, and task assignation, enabling the company to follow the individual path of workers, limiting the discretional decisions of foremen and supervisors. This individuality implied monitoring the construction of a body of subjects on a multinational level. Through the information produced by the labor department in the several divisions, the officers of the holding could follow and track problems with the workforce and control the appearance of organizers among them on a global scale.[56]

As this chapter has illustrated, the construction of international workers' bodies during the 1920s was a growing concern for the firm. The US Company required a homogenization of the work processes to maintain profitability on an industrial scale. The standardization of the new extraction technologies generated, on one side, a homogeneous work structure in the different divisions that required large amounts of unskilled workers in the underground and a powerful elite of skilled workers on the surface. During the first decades of operation, the firm exploited the divisions between the underground and surface workers by adding ethnic barriers among workers. Nonetheless, as mechanics and refinery workers took consciousness of their strategic position, and challenged the definition of the corporate body, the firm's managers tried to govern over the individual workers. This new form of corporate sovereignty over the international body of workers tried to directly control the individual workers, their bodies, their organizations, and their living spaces. The company officers started some programs of lateral integration of the workforce in the distinct divisions through a common experience, the only one available for blue-collar workers of the firm: physical trauma.

The experience of trauma in mining was a common feature of the new industrialized mining, and the firm had established growing programs for safety in the distinct divisions. As an alternative to the formation of unions, the company created welfare meetings. These presented reports about the company's accidents and offered opportunities for workers to talk with the firm's officers and representatives of the welfare committees. These committees were the authorities in the division that authorized payments to wounded employees and the employees benefits fund's compensations. Additionally, the divisions organized safety committees in the firm's different productive units, reflecting the Leviathan's hierarchical ladder. The Shop Safety Committee, for instance, was dedicated to workers and was the authority that could suggest changes in the procedures to the Foremen's State & Welfare Committee.[57]

The new policies of the labor departments functioned as a labor governance for the firm: to register, control, track, and, finally, make retribution. The labor departments dealt with the organization of safety teams in the mines. Early in the 1920s, the divisions widely used the manuals of first responders of the US Bureau of Mines, even in the Mexican operations. In every division, they nominated safety delegates, organized in crews, and had to train more than twenty hours with standardized equipment. The crews designated for every subsidiary and productive units were continuously trained, updated with alternative methods, and took part in local competitions. In Utah, the firm organized the meeting and annual demonstration for first aid in the Utah state fair, attracting attendance by the governor, the state's secretary, and the general prosecutor. The mining, metallurgic, and municipal and state service teams solved unexpected problems about the wounded in the contest, applying bandages and tourniquets. In Alaska, the firm organized an annual contest of first aid in the Chatanika mining camp, where the small dredging companies of the area competed with the several mining camps of the Fairbanks Exploration Company. The participants had to solve problems, including malfunctions of the pumping mechanisms to rolling cars and electrocutions.[58] The team of the Mexican division was by far the most celebrated. It was considered the best in the country and often won the national safety contest. This display of the firm's safety teams did not only exist on a local and national level, but also revealed the international operation of the firm. In 1923, during the international safety meeting of the US Bureau of Mines celebrated in Salt Lake City, the Mexican division of the US Company won the binational championship in first aid rescue teams.[59]

The growing visibility of the physical dangers of the mining activities and the construction of a generalized policy around it revealed only one effect of the trauma of mining on the workers' body. Under the program Safety First, the company installed signs in the mines telling the regulations of safe work inside the company's operation and regularly distributed stories about how the laziness, carelessness, or avarice of miners could lead to death or injuries. Accidents were, in the end, attributed to the negligence of the employees, either the injured or one of their peers. In other cases, they were simply considered a natural and accepted risk of the miners in the execution of their work. By law, the firm had to maintain certain standards of operation to assure the workers' safety, but filing legal claims against the corporation was useless. The legal team usually argued that the firm was incorporated in Maine and asked for a replacement of the complaint, delaying the process and undermining the capacity of the families of the deceased or wounded to continue the legal action.[60] In short, the prevention of accidents did not appear as a concession of the firm's technicians, but as part of the educational project of the firm to form a responsible and efficient body of workers. The installation of hospitals or resident doctors in all the divisions of the firm, from Midvale to Pachuca, from Northern California to southern Utah, and from East Chicago to Fairbanks, appeared as a therapeutic intervention of the company over the body of workers, a royal touch against the perils of incompetence, avarice, and negligence.

By the end of the 1930s, another growing concern appeared over workers' bodies: the effect of the dust particles on their pulmonary function. In 1933, Blamey Stevens, manager of the Lane Rincon Mines Inc. in Temascaltepec, Estado de Mexico, started a campaign among its workers. He opposed the introduction of union organizers into his mines and attributed to them a demand for pneumatic drillers. In the flyer that he distributed, which reached national relevance, he equated the *contamination* of radical ideas among workers with the contamination of the worker's body by industrial mining. Industrial unions and industrial corporations were both threats to his traditional paternalistic company. He insisted that he only allowed pneumatic drilling in the lower levels, as he believed the mechanized methods gave the workers only eight years of useful life. On the flyer, heavily circulated among miners, he enumerated how industrial mining destroyed physically and spiritually the body of workers, and made the promise of returning to agricultural life vanish: "Finally, you take your compensation and start to work on your land. And you are surprised how fast the money runs dry:

it was so much at the beginning! Even worse, you do not have even the strength to work the ground... All the money is gone, and your family needs charity." After the miner's death, in Stevens's words, industrial mining's curse followed the worker to his household. "Because your silicosis has transformed into tuberculosis, which you have transmitted to your children, and in that way, all your family is ruined."[61] Stevens's strange anti-unionization campaign conveyed a red alarm to the managers of the multinational firms operating in the country, and the flyer reached the headquarters of the firm in Boston. It rang an alarm bell for the top officials, as it fostered a growing awareness of industrial diseases in the body of workers.

Later that year, on October 17, 1933, Felipe Garcia Contreras, whom we met earlier, took a small card with some of his data, name, and age, numbered 267 1 800 11 45, from the labor department clerk. The next step in the application process was to go to the Real del Monte hospital, forty minutes on foot, to take a medical examination. There, the officers weighed and measured him: he weighed 121 pounds and stood 5 feet, 2 inches tall. He was not big, but only a couple of inches under the average. The radiologist performed the most complex and most important physical test: an X-ray of the chest. Along with the rest of the applicants that day, they sent home Felipe to wait for the results in Loreto's offices.

Intervention into the miners' body in the hospital usually had to do with the treatment of trauma resulting from the production process. Nevertheless, the role of the examination of the medical team for applicants was to evaluate the laborer's fitness to be incorporated into the sovereign body of the firm. As such, the selection reflected the search for the ideal individual body of the worker. The US Company rejected around half of the job seekers based on their pulmonary condition, besides those rejected based on previous jobs or union activity. Most of the refusals regarding respiratory conditions were old-time miners, who had left other jobs in the past, or other mining operations. The radiologist examined the display of Felipe's lungs and saw minerals imprinted on them. He punched the applicant's card, "Not apt for work." The clerks at the archive filed the report of the applicant 17386T for legal reasons, but the recruits themselves, accepted or rejected, had no access to their medical records. Felipe never went inside the mountain, as the mountain was already too much inside of him.

CHAPTER THREE

Cosmopolitan Diseases

If the medicine is not effective, [the scrofulous] should go to the King.
They may be touched by him, and be blessed; because the illness is called
the disease of the kingdom, and the King of England has the most noble
touch. Finally, if this is not sufficient, turn to the surgeon.

—John of Gaddesden, *Rosa Medicinae*, 1307,
quoted in Bloch, *Rois Thaumaturgues*

DEPARTMENT OF GEOLOGY, BOSTON TECH

In September 1913, Professor Waldemar Lindgren gave the last touches
to the Cortez Associated Mines report. That company, incorporated in
New Jersey, had its headquarters in Boston, near Lindgren's offices, at
the Department of Geology of Boston Tech. The report was the result
of his travel, at company's expense, to examine the potentialities of the
districts of Zimapán and Jacala in the northwestern limit of Hidalgo.[1]

The method Lindgren followed in all his reports was to write the
history of the space. Like any good story, Lindgren's history started in
the future, looking at the potential developments of the railroad line
from Pachuca to Zimapán and the potentialities of accessing the mineral refining plants in Hidalgo or Guanajuato. Then, he recalled the
district's mining history, marked by the colonial exploitation of the
lead-silver mine of El Carrizal. He described the superficial traits of the
ground revealed in the Infiernillo cliff. He unveiled the different eras
by describing the rocks from the surface down. The Moctezuma River
had eroded some rocks and revealed others that could be read in a multiplicity of layers and breaks in the earth's crust. The area was covered
by lava flows and a tuffaceous bed of the Tertiary age, but the river had
eroded the rocks and showed more ancient phenomena, like a Cretaceous blue limestone and a black cherty limestone of the Lower Cretaceous. Some intrusions broke these different layers of sediment of
volcanic or igneous rocks: a light-colored granular or coarse porphyry.

While the sedimentary beds of limestone were aligned horizontally, the reddish porphyry had a diagonal or vertical distribution, with inconsistent contact points with the rest of the rocks. By looking at the abyssal time of these signs, Lindgren came back to the future, predicting metallic mineralization in the area, with some copper and iron deposits but also some galena and pyrite concentrations, containing lead, zinc, silver, and gold sulfides.[2]

Scientists had looked for signs of underground matter on the surface for centuries. In the sixteenth century, the alchemist and physician Theophrastus Von Hohenheim, known as Paracelsus, compared miners' works with physicians' in interpreting symptoms. Signs corresponded with reality since reality was the sign of the essence: "No mountain, no cliff, is so vast as to hide or conceal what is in it from the eyes of man; it is revealed to him by corresponding signs." Miners could discover everything inside mountains by external signs that corresponded with the underground, and physicians could identify the signs of disease. If the miner crushing and refining ore helped organize matter into its most perfect state, the surgeon did the same, purifying the human body and making it as indestructible as gold.[3]

During the eighteenth and nineteenth centuries, the reading of signs in the mountains by geologists revealed the origin of the earth and men's place on it. The order and position of rocks and fossils was at the center of the debate on the history of the earth, between gradual and catastrophist processes, and between the forces of deluges and volcanoes in the configuration of the crust. The debate over rocks was vital to the creationism of Georges Cuvier and the evolutionary theory of Charles Darwin and Alfred Wallace.[4] The earth's signs revealed the history of nature but also allowed geologists to locate valuable minerals. Most of the geologists who debated the origin and temporality of fossils, such as Gottlieb Werner, James Hutton, and Charles Lyell, either consulted for mining companies or surveyed new territories for states.[5] In short, mining and geology experts were interested in the ground not only as a natural force to understand, or as a mystery, but also as an imperfect resource to locate and extract: a mineral body to conquer.

Lindgren was an expert on the interaction of water with volcanic activity. Born in Sweden, he studied at the prestigious Freiberg School of Mines, and he had worked in the United States Geological Survey since 1884, which played a central part in the United States' mineral frontier at the end of the nineteenth century. He surveyed for gold in the Blue Mountains of Oregon, underground water in the Molokai island of Hawaii, copper in Clifton Morenci, Arizona, and silver in Idaho.[6] In 1912 Lindgren was appointed head of the Department of Geology at MIT. The transition from a government office in Washington to an academic office in Boston also implied being a consultant for the private sector, as the networks of firms, academics, and professional organizations interlapped. The firm that employed Lindgren to examine the deposits in Zimapán, the Cortez Associated Mines, focused on the location, development, and sale of promising claims to big firms. One of the obvious choices in the area was the United States Company, which pursued an aggressive policy of expansion led by the exploration department head, Sidney J. Jennings. Lindgren knew Jennings well, as they both formed part of the council of vice presidents of the American Institute of Mining and Metallurgical Engineers.[7] Moreover, another geology professor at MIT, William Otis Crosby, had consulted the Bingham Mines Company just before its absorption by the US Company.[8]

The connections of Lindgren with mining firms were not rare, as MIT's geology and mining department relied heavily on the professors' consulting jobs and links to corporations. In the coming years, Lindgren increased his relations as a consultant with mining firms, asking for permits to leave his administrative obligations to evaluate or survey new properties. His consulting jobs were crucial in the expansion of US interests across the hemisphere. In 1916 he evaluated the iron deposits in Firmeza and Daiquiri, Cuba; one year later, he surveyed the Braden and Chuquicamata mines in Chile; and three years later, he examined the tin deposits in Bolivia.[9] Soon after the mechanics strike in the US Company properties in Mammoth, California, the firm required Lindgren's services to evaluate its properties, which had been explored by a small crew who had projected increased reserves to 25,000 tons. Lindgren pointed out the existence of at least 185,000 tons of ore reserves belonging to the firm and predicted that that amount might duplicate after further exploration. As a result, the firm increased its development in the area and resumed the mines and smelter operations in 1918.[10]

If the geologist extended the corporate sovereignty over mineral bodies on the ground, the territoriality of mining firms shaped the way scientists understood those mineral bodies. Lindgren's practical approach to locating minerals revolutionized the exercise of economic geology, combining the photographic and chemical examination of rocks with the topographic and aerial survey of the territory. Lindgren's widely used textbook, *Mineral Deposits,* reflected a catalog of different rocks, occurrences, traits, and signs of metals in different parts of the world. The contamination and interactions of different kinds of minerals were the basis of Lindgren's new approach: "The student who is familiar only with the fresh, unaltered specimens finds himself in the midst of puzzling and strange types that he is unable to classify with certainty."[11] The practical character of Lindgren's work and classes shaped a generation of geologists and mining engineers with subsequent experience in the field not only as consultants but also as managers and surveyors in both public and private enterprises. In the 1920s, Lindgren's former field assistant, who had experience in mines in China, Mexico, and South Africa, became increasingly relevant in the national political scene: Herbert Hoover.

This chapter argues that the dynamics of consultancy did not only imply a relationship between the corporations and the expertise, but also revealed signs of a more extensive correspondence between the planning that was done inside and outside big organizations. The consultants associated with the US Company, the protagonists of this chapter, researched the behavior of two elements: sulfur and silica, and their relationship with mineral and human bodies. Geologists and metallurgists working for the corporation were concerned with identifying and extracting materials, designing prospecting technologies, and separating their components. Physicians used materials to identify diseases and cure them, and the natural body's contamination of the human body. Some of these experts were foreigners to the corporate body, but they appeared as the alchemists transforming elements from the outside into poison or therapy on the inside. Their story and expertise were as important to the firm as the managerial capabilities of the officers. In the same way, the corporate body was essential for constructing their expertise and their academic institutions. This history of interaction between chaos and organization, between metals and sulfur, took place in the interior of the human body and the exterior of the globe and left its prints in the correspondence within every part of the corporate Leviathan.

METALLURGIC LABORATORY, MIT

In the first days of 1917, the students of mining and metallurgy at MIT started populating new lab installations in the new campus designed by Rockefeller's architect, William Bosworth. In June 1916, MIT moved to its current location in Cambridge, but during the fall semester, the students in the course in mining and metallurgy continued to do their experiments in the installations of the basement of the old Rogers Building, in the original Boston Tech building located in the Bay. Little by little, the furnaces and machines from the old laboratory populated the surface installations in Building 8, and new machines arrived as gifts from technological corporations. The Dwight-Lloyd Sintering Co. of New York City donated a Dwight-Lloyd sintering machine and Utley Wedge, from the Furnace Patent Company, donated a six-hearth roasting kiln: both pieces of equipment were designed to extract metals from sulfide ores.[12]

Despite the founders' large projects, in the first years of the twentieth century the school remained a small and struggling polytechnic institute. The MIT had, up until 1913, just a tenth of Cornell's annual income, and multiple times had tried to merge with Harvard. Several inside groups disputed the direction that the institute should take. Arthur Amos Noyes, professor in chemistry and acting president of the institute between 1907 and 1909, tried to push the MIT to become a liberal arts yet science-oriented school and created the Research Laboratory of Physical Chemistry. Others defended the institute's practical character, including Professor of Chemistry William Hultz Walker, and tried to increase the curriculum's relationship to actual engineering problems.[13] Walker founded the Laboratory on Applied Chemistry and reinvigorated the program in industrial chemistry, organizing it around the different techniques used for separating materials. The scientific and practical approach of Walker did not privilege obtaining knowledge, rather the transformation of the world. He considered himself one of the heirs of Paracelsus's arts: "The captains of industry and their army of co-workers are still alchemists at heart; they still strive to transmute the base materials of the earth into gold."[14]

Walker gained relevance in the fight for the control of the chemistry department after Noyes's presidency, as a wave of alumni pushed for the institution to build relationships with firms. Walker was an industry favorite, and his corporate connections were determinant in the funding for the new campus in Cambridge, made by gifts of the tycoons of Dupont and Kodak. Walker himself had an ongoing relationship with Arthur Little, one of the biggest engineering consulting firms. This pragmatism informed the organization of the School of Chemical Engineering Practice in February 1917. The project implied that a select group of students learned the chemical processes in the industry in the plants of the Eastern Manufacturing Company, the New England Gas and Coke Company, the Carborundum Company at Niagara Falls, the American Synthetic Color Company, and the Atlas Portland Cement Company. Other professors followed the path of integrating research with big firms, especially in the mining department, which shared a building with the chemical engineering program.[15]

In April 1917, Heinrich Hoffman, head of the Department of Metallurgy and Mining, signed an agreement with the United States Company. The corporation provided an annual allowance of $3,000 and two research scholarships every year in their labs. In return, the firm had at its disposal all laboratory facilities of the institute, the library, and the faculty, and could install its Central Research Laboratory offices at the metallurgical department. The management of the US Company could name the person in charge of the laboratories, who would be paid as a research associate and was in charge of the publication of results.[16]

Mining and smelting firms had established even bigger laboratories within multiple subsidiaries during 1916, but the lab of the US Company was a pioneer in formalizing its relations with academic institutions. The central laboratory was about to take control of the research officials' activities in the rest of the company's divisions and communicate with the corporate headquarters in Boston for key findings. Research at MIT helped coordinate the other labs and benefited from practical results in the rest of the labs. The engineers in the Mexican division of the US Company focused on experimentation with new explosives, as the previous years there had been a huge scarcity of TNT and powder. In the next years, the division followed an ambitious program of using liquid oxygen combined with liquid nitrogen to substitute for dynamite through a German patent.[17] The metallurgy department of the US Metals Company, the US Company's subsidiary in Chrome, New Jersey, had a research staff of forty-four engineers and scientists, led by H. D. Greenwood and W. C. Smith. The research in those facilities specialized in the electrolytic separation of sulfides to refine copper, lead, and silver concentrates. Finally, the Midvale plant as with the other divisions, had some of their engineers focused on research. Led by Galen H. Clevenger, the division focused on volatilization during the smelting process. It also operated an experimental baghouse and an experimental farm concentrated in the investigation on a subproduct of sulfides treatment: the expulsion of sulfur dioxide into the environment.[18]

Salt Lake City was a center of research for several companies. The Kennecott Research Center controlled and expanded its research on its four western mining divisions. The Kennecott's director of research coordinated from the central laboratory experiments to recover copper, gold, and molybdenum in the metallic giant's different divisions. The American Smelting Company also installed its central laboratory in their plant in Murray, in the Salt Lake Valley. The firm had an office specializing in research on the effects of sulfur dioxide and heavy metals expulsion into the atmosphere. Utah Copper, and later Kennecott, located a laboratory in its smelter in Garfield, Anaconda, and its central labs in Butte, Montana. However, research and development was coordinated with similar departments in every division. Finally, the research departments of the various companies reached out to universities in order to pursue their objectives. The US Company, Kennecott, and Anaconda offered similar fellowships to engineering and mining students at the University of Utah.[19]

The day-to-day activities of the central laboratory of the US Company at MIT were mainly to test different assays of the exploration subsidiary of the firm and a proportion of the ores from the divisions in Utah and Mexico. The laboratory solved some ore dressing and extraction problems in the firm's metallurgical plants, allowing the company to locate the distinct elements in the metal sulfides it treated and discover the best ways to extract the metals from them. The Cambridge location was key for the holding officials, as they could use it to control the expansion of the subsidiaries, homogenize the treatment methods of different ores, and keep the strategic decisions in Boston.[20] The alliance with the firm also allowed for some basic research at the institute. In just the first year of operation, the scientists in Cambridge investigated elimination of tellurium from the anode mud of the Betts process, developed a new process for the production of lead arsenate, produced cadmium pigments, recovered bismuth from flue dust, determined bismuth in lead slags, and detected some rare metals in meteorites.[21]

The research department of the US Company combined the typical research on roasting and furnaces with chemical processes to separate ores. Beginning in 1918, the corporation started experiments using the flotation method to separate the minerals after crushing. The traditional process of concentration was based on the different weights of the particles in the crushed ore. Separation tables vibrated, expelling most of the lighter components, usually silica and dust, and conserved the metallic particles, which are naturally heavier. Although effective, the method wasted large quantities of low-grade ores, and the metallic firms experimented with the froth flotation method for mineral sulfides. In froth flotation, a particular chemical component is added to the crushed ore, which reacts with the sulfides and turns them hydrophobic. After the reaction, the pulp is exposed to air and bubbles, and the newly hydrophobic metallic sulfides rise to the surface as concentrates. The firm started the research in 1918 on the different ores and components, for ores containing lead, silver, and copper, and installed units in the distinct divisions. In addition to evaluating the different concentrations and combinations of the ore, the research lab at MIT had to evaluate the effect of altitude in the reactions conducted by different scientists working in the field, and the different treatments of the sulfides in the roasting, cyanidation, or smelting processes.[22] During those endothermic reactions, the liberation of gases provoked the appearance of small dust particles that contained metals and sulfides that had a commercial value in precious metals, like silver, but were highly harmful in lead and copper sulfides. In those years, the research teams in the plants at Midvale and Chrome investigated the recovery of metallic and sulfide particles during the smelting and refining processes.[23]

In 1920, with an increased volume of lab operations, the US Company started using a second room at MIT for the research department offices.[24] William Walker tried to extend the partnership to other firms through a technology plan. Pulling away from the philanthropic model of financing, the new structure would require an agreement between an industrial organization and the institute: the firm paid an annual fee, while the university provided the technicians of the firm with training, suggested staffers, and allowed the technicians of the companies to use the labs and libraries of the institute. Finally, the faculty of MIT would consult and research on topics related to the firms' operations.[25] The initiative was, in short, an extension of the agreement with the US Company and was rapidly supported by the industrial community. "Carried to its conclusion, the Technology Plan will make the Institute of Technology the greatest consulting body in the World," declared Walker on January 10, 1920, in the plan's inauguration. The program, part of the funding campaign of the institute to increase the endowment by $8 million, was a tremendous success, and over 150 firms signed agreements with Walker as the head of the newly founded Division of Industrial Cooperation and Research.[26]

The emergence of MIT as a consulting body was not simply about building a relationship with corporate entities but also about building warfare knowledge for the state. The institute's expertise had been instrumental in the foundation, in 1918, of the Chemical Service Section of the army. Van H. Manning, the director of the Bureau of Mines of the Department of the Interior, had been commissioned to develop research on the Germans' chemical weapons. The institute's know-how on poisonous gases in the mining and metallurgical spheres, especially sulfides, was crucial to the war effort. William Walker became commanding officer of the Chemical Warfare Service, and one of the priorities was the research on dichloroethane sulfide, a chemical compound used by the Germans to attack the eyes and respiratory system: it was commonly known as mustard gas.[27] Walker had to declare, in 1919, some business interests in corporations that provided inputs for the army during the war, as a procedure to reduce conflict of interest in his work as a defense consultant. As head of the Chemical Warfare Unit after the war, Walker acknowledged that he had shares of some mining and metallurgical firms that provided inputs to the army: the United States Company was at the top of the list.[28]

The circulation of knowledge about poisonous gases between the corporate bodies and nation-states continued in the following years, relying on the same mining and metallurgical technicians. By the end of the First World War, the institute applied some knowledge in materials produced in chemistry and metallurgy labs to defense purposes. On February 1, 1922, the acting president of MIT sent an application to the adjutant general of the United States Army to establish, in the labs of the university, a Chemical Warfare Service Unit. The project proposed to include general training in chemical warfare to all the students of chemical engineering, and offered a summer camp to other Reserve Officers' Training Corps units. For the faculty at MIT, the proposal was a continuation of their work during the war and their previous research on chemical components. While the project attracted the army's interest, the request was put on hold, given the anti-chemical weapons rules of the Conference of Limitation of Armament, which limited the use and development of chemical weapons.[29] Nevertheless, the firm's officials believed that the use of sulfides were necessary for sanitizing the workers' bodies.

TROPICAL HEALTH DEPARTMENT, HARVARD UNIVERSITY

In the first days of January 1916, the vice president and second vice president of the United States Company, Frederick Lyon and C. W. Van Law, wrote to the office of Richard P. Strong, professor of tropical health at Harvard Medical School. The previous year had been the worst in the Mexican division's operation, with the Pachuca and Real del Monte plants operating at between 50 percent and 80 percent of their capacity, as the *convencionistas* and *constitucionalistas* forces were disputing control over central Mexico. Communication was difficult, the production chain of the company was intermittent, and the war in Europe made supplies like explosives and chemical compounds scarce. On top of everything, in the second half of the year, the camps of the company experienced a massive typhus outbreak.[30]

Ten years before, the US Company's managers had inaugurated the miners' hospital at Real del Monte. The hospital was the first public event for the new managers after acquiring it from the former Mexican owners. The firm assigned an American doctor and nurses, and the hospital staff was on the top of the hierarchy at the division. The doctor and the geologist of the Compañía de Real del Monte had the best remunerations in the subsidiary, only behind the general manager and the superintendent. The first chief of the medical department of the firm, once it restarted operations years later, was E. G. English. He had met Van Law, one of the pioneers in the cyanidation process applied to gold sulfides, when both worked for the Guanajuato Reduction and Mines Company. In 1909, he followed him to Pachuca and equipped the hospital with state-of-the-art technology for treating trauma, the only consequence that they recognized mining had in the human body.[31]

When the typhus epidemic reached the mining camps in Mexico, neither Dr. English nor the Mexican doctors that assisted him had the appropriate resources to control the spread. Large human concentrations and poor hygienic conditions were the key factors that contributed to the propagation of the disease, characterized by abdominal pain, rashes, and extreme fevers. Between 1912 and 1915, the prevalence of typhus grew constantly, but it reached epidemic levels in the second half of 1915. The dispute for strategic points in the revolutionary war and the drought of that year catalyzed people's extraordinary migration into the main cities. In Mexico City, around 50 percent of the infected were military personnel, and the rest of the sick were concentrated in prisons, juvenile correctional facilities, slums, churches, and hospitals. More than six thousand people fell sick between August and December 1915 in Mexico City, with a mortality rate of around 20 percent.[32] By then, the US Company employed around four thousand workers in Pachuca, 10 percent of the area's total population. The focus of infection in Hidalgo was, in those moments, the housing and working areas of the mining firm.[33]

The meeting between Lyon, Van Law, and Richard Pearson Strong in Boston, Massachusetts, tried to tackle the crisis in the mining camp with the help of Strong's expertise, acquired in Europe the year before. In 1914, the Rockefeller Foundation organized the American Ambulance Hospital in Paris. The first secretary of the philanthropic organization, Jerome Davis Greene, had been secretary of Harvard, and for the first expedition to the European war zone, the foundation recruited most of its volunteers from Cambridge, Massachusetts. As part of this delegation, Strong arrived in Paris at the end of 1914, and he was immediately reassigned as the American Red Cross Sanitary Commission director, in charge of battling the epidemic of typhus in Serbia that threatened to spread to the rest of the Allies. The typhus outbreak worsened in the first months of 1915, and the number of lethal victims increased to 150,000. Upon his arrival to the East in April, Strong organized the International Health Commission to coordinate containment measures and incorporated the techniques of other imperial experiments. The French delegation, led by the bacteriologist Charles Nicolle, proposed a revolutionary theory of disease transmission, developed in Tunis: the disease was transmitted by lice. Nicolle went to Paris to instruct the French delegates, and his colleague, Ernest Conseil, and pupil, Georges Blanc, were transferred to Serbia.[34]

At least a dozen Harvard professors and graduates, some of whom had worked under General Gorgas in Panama, arrived in Serbia in the next months. The commission's first step was to disinfect and fumigate the hospitals, the jail, the barracks, and some stables filled with military personnel. In Strong's manual, typhus was a "cosmopolitan disease," an infection not particular to any single country, and one that appeared in most cities around the world as a product of human communications. He had gained knowledge of epidemics during his intervention in Manchuria in 1911, when he was in charge of containing the pneumonic plague in the Chinese territory occupied by the Russian Empire.[35] In Strong's protocol, the doctors and patients at the hospitals were stripped, their hair and nails clipped, and their bodies bathed while the personnel steamed the clothes and linen in order to sterilize them. The spaces were then cleaned and fumigated with elemental sulfur, which was burned to liberate sulfur dioxide into the rooms and kill the bedbugs and lice. Sulfur was the main supply material needed for the campaign, and a single hospital needed somewhere between 1,600 and 35,000 pounds of the yellow powder. However, sanitation also required an enormous amount of labor. The physicians and nurses wore louse-proof suits, rubber gloves, and masks to interact with the sick and used Austrian prisoners to care for most hospitals and wards' work. "Nearly all the prisoners had had typhus, and a very large proportion had died of it," described George Shattuck, one volunteer from Harvard. "Being immune to typhus from having had the disease, it was not necessary to take precautions to protect them."[36]

The commission successfully controlled the outbreak, and most of the crew members abandoned the area at the beginning of August, just when the typhus outbreak began in Mexico. The team's success in Serbia demonstrated the mechanisms of transmission and the measures to control and moderate the spread of the disease. Strong's work drew the attention of the upper management of the US Company. At the time of the interview with Frederick Lyon, vice president of the multinational corporation, in 1916, Strong was still weak from his European trip. He had suffered from malaria on his way back from Serbia and did not volunteer to personally go to the Mexican mining district. He offered, nonetheless, two useful contacts to the firm. He suggested the services of the physician Henry E. Berger, a student of public health at MIT who had been part of the supplemental party of sanitary inspectors in Serbia the year before. He also telephoned New York, searching for the services of Victor Heiser, from the Rockefeller Foundation.[37]

It was not rare that corporations reached out to Strong for advice on medical issues in their tropical divisions and that Strong recommended students such as Berger to take care of specific projects. The program in tropical health at Harvard was founded in 1913 to formally train government officials in the country's new colonial properties, especially Puerto Rico and Panama. The project, led by Strong, required the approval of the surgeon general, the US Army, and the secretary of war, and right away established connections with multinational firms. At the same time, Strong was appointed the director of laboratories for the United Fruit Company, as part of the new department's funding strategy. In many ways this arrangement was a mirror of the formal alliances of mining firms with geology departments in Boston. In the summer of 1913, Strong organized the first expedition of students from the tropical health department to the tropics, visiting the district of Junin, Peru, where the US Company and the Cerro de Pasco Mining Company dominated the mining extraction.[38] During the trip, true to his practices, Strong forcefully inoculated a Chilean patient of Lima's mental asylum (Manicomio del Cercado) with *verruga peruviana*, trying to disprove its relationship with the deadly Oroya fever common in the mining camps.[39] In the coming years, the tropical health department's formal alliances implied that, most times, the hiring of doctors for multinational mining firms, both American and Latin American with US training, was done in the offices of Strong.[40]

The suggestion of Victor Heiser to take charge of the project combined corporate, imperial, and philanthropic projects in the Global South. Strong and Heiser had met in the Philippines years before, when Strong worked at the Bureau of Science in Manila, and Heiser was the director of health in the territory. During those years Strong was focused on experimenting with inmates of the prison Bilibid in Manila, a panopticon jail built by the Spanish Rule fifty years before. In 1905 alone, Strong had forcefully injected his anticholera vaccine in 1,662 prisoners, and in

November 1906 he tried a different formula in twenty-four inmates over the age of seventeen, without prior knowledge and dominating them with the help of the guards. The serum was contaminated with a virulent strain of bubonic plague, and in the next weeks, all the test subjects sickened, and thirteen died.[41] After a short investigation by the authorities, Strong dropped his cholera research, but he continued experimentation with the prisoners in Manila. In 1910 he led another experiment with twenty-nine inmates sentenced to death, in order to research the effects of nourishment in developing beriberi. They isolated them into groups, restricted their diet participants to different degrees, and fed them different kinds of rice for four months. All prisoners lost weight, from five pounds to twenty pounds; fifteen developed the disease; and one died during the protocol.[42]

Similar to Strong, Heiser believed in the value of segregation and forced practices to control diseases in the tropical bodies. Like many, Heiser believed that the habits and physiology of Asians, combined with the Spanish colonizers' laziness in controlling them, were the causes of the disease on the island. He believed that the islands' population ranged from "dependents" to "wilds" and lamented the transition to a civil government, and the impossibility of implementing harsher control strategies.[43] Segregation was the principal component of Heiser's career in the Philippines, where he was in charge of facing the outbreaks of plague, yaws, cholera, tuberculosis, and beriberi. He was responsible for constructing the Philippine General Hospital, mass hospitals for infectious diseases, mental asylums, and the confinement of leper patients in the Culion Leper Colony, commonly known as the Island of No Return. He was, more importantly, one of the coordinators of the mission against the plague in Manchuria and was therefore familiarized with the use of sulfur in the methods of Richard Strong.[44]

At the same time, the Rockefeller Foundation hired Heiser to oversee the organization's fellowship program. The philanthropic organization was incorporated in New York in 1913 as a conciliatory mechanism between business and society independent of the sovereignty of the state; it was a formalization of the tycoon's rural health campaigns in the American South. The foundation expanded its operations into Latin America and the Caribbean in the following years, focusing on research on malaria, yellow fever, and hookworm. Heiser served as the link between the philanthropic organization, universities, and multi-national firms that expanded to the south of the continent. Following that endeavor, he built an alliance with the tropical health department at Harvard and the United Fruit Company, both under the auspices of Richard Strong.[45]

Many roads of expert knowledge from the first two decades of the twentieth century converged in the meeting of Lyon with Strong in the first days of 1916. When Strong tried to reach Heiser after his meeting with the US Company's officials, he learned that his old collaborator was already in Mexico as part of the Rockefeller Foundation mission to implement a hygienic intervention in the country. The foundation negotiated with the de facto government, led by Venustiano Carranza. As US troops and carrancista forces had clashed less than two years before in Veracruz, the task was difficult, so the foundation asked for the State Department's services to mediate. The previous experience of Heiser as a government official allowed him to access the country as an emissary.[46] He carried with him a succession of credentials as imperial officer, hygienist, philanthropist, physician, and corporate consultant. The main method used to battle the outbreak was not new for him, nor for some of the medical officers in the mining camps. Fumigation with sulfur was not completely new for the hygienists in Mexico, as in 1892, municipal officials had implemented it in the mining town of Zacatecas.[47] The experts controlled the outbreak within a month.

Strong's help to the US Company during 1919 was not a coincidence, as mining firms and their expansion were vital to the consolidation of the tropical health programs at Harvard. In the following years, the mining corporations operating in the western and southern territories, and financially in New York, became more involved with the cosmopolitan hygienic and health institutions. One year later, in 1917, Henry Pomeroy Davison, president of the trusts that controlled the US Company stock, the Bankers Trust and Guaranty Trust, was appointed chairman of the War Council of the American Red Cross. After the end of the war, Davison coordinated the International Committee of the Red Cross, continuing as chairman until his death.[48] In the 1920s, the Rockefeller Foundation established permanent operations in Mexico, with a decades-long campaign against yellow fever in coordination with the Mexican government.[49]

To be sure, the complex circulation of fumigation techniques using sulfur from Manchuria, Tunis, and Serbia into Pachuca was not an accident. Strong's interview with Lyon was only an episode in his life and the general operation of the US Company. However, it revealed structural connections consisting of a tight circuit of knowledge formed by international philanthropic organizations, public health schools, national governments, and multinational corporations. Most health officials moved freely between public institutions, universities, or multinational firms, and these organizational networks implemented undifferentiated methods of control over workers, colonized people, war combatants, and prisoners alike. Experts transitioned between several flags, but all coincided on the use of sulfur as a disinfecting agent. The convergence of the international philanthropic organization, the public health school, the federal governments, and the multinational corporation in the interview between Strong and Lyon concerned the extraction and use of sulfur and its application into the workers' bodies. At the same time, another group of experts was investigating the role of sulfur not as a disinfectant nor as a container of metallic richness, but as an industrial poison.

HULL HOUSE, CHICAGO

On February 12, 1913, Dr. Alice Hamilton wrote a letter to the Bureau of Labor commissioner, from the Department of Commerce. It was a rather annoying matter. Five days before, she had received a communication at the Hull House, where she and two of her sisters lived in the south of Chicago. The commissioner sent her a copy of a statement published in the *Ohio State Journal,* summarizing a meeting in the state board of health. In the statement, Dr. H. T. Sutton had called the report of Hamilton on the potteries and tiles in Zanesville, made for the federal Department of Commerce, "a malicious and slanderous report, or an erroneous one." Dr. Sutton declared that the levels of lead in the district were minimal and harmless to the workers. "But this woman"—he continued—"discovered that almost the entire force of employees was suffering from lead poisoning in some form or the other." In her answer, Hamilton seemed remarkably vexed. Sutton, she responded, was a member of the state board of health but at the same time he was employed by the American Encaustic Tile Works of Zanesville. "He has seen a good deal of lead colic and is often called to treat cases of hysterical convulsions in girls, but he does not know whether these girls come from the glaze room or not. He never asked, thinking it was always hysteria."[50]

Hamilton was engaged in a large survey of lead smelters. She had surveyed eight establishments around the East and Midwest, and she had five more on the list: all of them were plants of ASARCO, Anaconda, and the US Company. The new study was very close to the scientist, both geographically and personally. She was the second of five siblings raised in a wealthy and educated family in Fort Wayne, Indiana. She attended the University of Michigan Medical School at Ann Arbor and later traveled, with her older sister Edith, to Germany, trying to attend the universities of Leipzig, Munich, and Berlin. After an unsatisfactory experience, due to women's restrictions in classes, the Hamilton sisters came back to the United States. Edith became a renowned and published classicist, while Alice spent time at Johns Hopkins University before accepting a position as professor of pathology at Northwestern Women's Medical School. That year, 1897, she also was accepted as a resident at the Hull House, where she lived for twenty-two years.[51]

Founded in 1889 by Jane Addams and Ellen Gates Starr, Hull House was a settlement of educated women. Alice and, some years later, her youngest sister Norah were volunteers who gave classes to low-income women and neighborhood kids. Hull House formed part of a growing network of social reformers who took part regularly in rallies and picket lines for workers in the area. In those first years of the twentieth century, she met a wide range of radicals who visited the house: Patrick Geddes, Bill Haywood, the Webbses, H. G. Wells, Nikolai Avksentiev, Peter Kropotkin, Emma Goldman, and Marie Sukloff, among others. Alice Hamilton remembered a young Upton Sinclair, resident of the University of Chicago Settlement House who was writing *The Jungle*, who told her that the socialists would inevitably win the presidency.[52]

In the first years of her residency, Hamilton opened a well-baby clinic and started some work on the hygiene of the working classes and its relationship with infectious diseases. Hamilton's initial approach analyzed the origin of infections among the habits of the poor, studying the typhoid outbreak of 1902.[53] Hamilton conducted further research on another common disease among workers in the following years: tuberculosis. The traditional approach of physicians and doctors in the workplaces was to reform laborers' practices, considered either hygienic, dangerous, or reckless, and to treat the trauma provoked that, they considered, was self-inflicted. Alice Hamilton had been part of that same wave in the first years of her career, but this time the physician focused not on the workers but on work. She measured the fatigue of the factory girls with an ergograph borrowed from the University of Chicago and related her findings to the disease's expression.[54]

The transition from pathogens to toxins in Hamilton's practice had everything to do with the region's working environment. In 1910, Professor of Sociology at the University of Chicago Charles Henderson convinced the governor of Illinois, Charles Deneen, to appoint an Occupational Disease Commission. It was the first of its kind in the United States and was composed of five physicians, an employer, two members of the state labor department, and Henderson. While others studied carbon monoxide, deafness, Caisson disease, and coal miner's nystagmus, Hamilton's field was lead. She discovered many industries that used the metal in some part of their production process and focused her attention on the less-evident symptoms related to poisoning, like dementia in very sick patients and anemia with minor levels of exposure. In 1911, the report had pushed the state to regulate the operations of some lead industries, although Hamilton considered them very insufficient.[55] While she was still working on the Illinois report, the Department of Commerce requested a general investigation of the lead smelting industry nationwide. The assignment marked the definitive transition from bacteriology to toxicology in her career: "I accepted the offer and never went back to the laboratory."[56]

Hamilton's research on lead smelting took four years, and, before concluding it, she published the report of lead used in the potteries industry that enraged Dr. Sutton. In the course of her work, she had dealt with more than her fair share of arrogant and incompetent male company doctors. Most of them had no notion of the plants' working process, believed that the workers absorbed lead through the skin or by eating it, and gave no attention to the enormous quantities of free heavy metals present in the air as dust. Hamilton contended that lead was absorbed more easily through the respiratory tract than through the digestive system, and that workers ingested lead far less often than thought by their physicians. Some company doctors provided the workers with rubber gloves or installed baths to clean the men after their shifts. One of those physicians, she recalled, recommended that the laborers scrub their fingernails to avoid lead poisoning. Most of them only recommended a heavy rotation of workers in the plants to prevent larger exposure and to avoid any legal responsibilities of long-term employment.[57]

In the coming years, Hamilton became familiar with metallurgy processes carried out in the plants nationwide, finding the same big firms in all the locations, especially the US Company and ASARCO. Both corporations had operations in the major areas Hamilton surveyed: Salt Lake City, south New Jersey, East Chicago, and the tri-state area between Missouri, Oklahoma, and Kansas—and each followed different protocols. The US Company used, she described, Midvale converters, a company patent of pot roasting, together with Dwight-Lloyd machines to prepare the ores for the furnaces. She reported that the dust in the ore's charge and discharge was the first exposure of the workers to lead intoxication. Next, the concentrated and pre-roasted ores, red hot, were dumped into a pot with a fake bottom, forcing the blast into the cake with the ore. Regularly, the workers had to open the furnace to poke the pot, which assured the precipitation of lead oxides. The discharge of the lead oxides was the second point of heavy exposure to the lead dust and fumes. The molten cake dropped from a height into big pieces, but further breaking of the red-hot ore was needed.[58] Hamilton observed "men running up into the cloud of dust and fumes to break with hoes the larger pieces and push them through the grating. These were the men who were to escape poisoning by brushing their nails."[59]

The same problems appeared in the feeding and tapping areas around the furnaces. As with the converters and sintering machines, the process of charging the furnaces with fine particles and discharging molten metal and slag liberated fine metal components into the air. Most of the gases went through a baghouse in the Midvale plant to collect the fumes' particles. Spraying the smoke during filtration was impossible, as lead sulfides would transform into sulfuric acid, so the entire process maintained the dust in the air for workers. The baghouse at Midvale experienced explosions at least once a week, and the turnover of workers was tremendous: the managers hired more than one hundred men per month for the operation of a plant that required only twelve employees. Besides exposure to lead sulfides, the plants of Midvale, Leadville, and East Chicago of the US Company also refined arsenic, which produced skin lacerations.[60]

The work of Hamilton was, in many ways, the complement of the work of the company's geologists in predicting the mineral content of ore bodies. Hamilton did not consult the color of the rocks or the sample's testing to predict the metals of the land. However, she examined the records of mining hospitals in several regions and looked for long-term disease marks—evidence of the presence of toxic mineral substances—in the miner's body. She also visited asylums, examining effects on the worker's body and mind. The patients had a look that combined premature aging with tiredness and confusion. They showed the known effects of lead poisoning: "lead colic," a generalized and erratic, spasmodic pain in the abdominal area; "lead line," the formation of a black lead sulfide line on the gum on the margin of the front teeth; and double wrist drop, as very exposed patients lost the ability to use their hands. Hamilton underlined the relevance of early disease symptoms, like anemia, and the long-term consequences over the vascular and cerebral systems. Plumbism adhered to the arteries, making them less flexible and provoking a continuous deterioration of the organs as they did not receive enough oxygenated blood. Dementia was one of the vascular consequences, but lead also had direct toxicity with brain cells. The metal in the spinal cord paralyzed different extremities, which caused lead palsy and originated the contractions that developed into colic.[61]

The report was well received by the bureau but had little effect on the industry, as the Department of Commerce had no power to interfere with the companies' relations with employees. The beginning of the war redirected Hamilton's investigations. A committed activist for international pacificism, she, along with Jane Addams, attended the International Congress of Women in The Hague in April 1915. That same year, she began to work on a report on the effects of industrial war over the human body, examining the manufacture of explosives. She admitted it was a highly depressing job: "It was not only the sight of men sickening and dying in the effort to produce something that would wound or kill other men like themselves, but it was also my helplessness to protect them against unnecessary dangers."[62] Nitroglycerine is a product made from the combination of glycerin, nitric acid, and sulfuric acid. There were some similarities in the liberation of sulfuric acid between the extraction of metals in the West and the fumigation of hospitals in Serbia. The factories had to roast the metallic sulfides, mainly pyrite, and then precipitate them and cool them with water, liberating large amounts of heat, sulfur, and nitrate oxides. Workers were susceptible to developing chronic pulmonary diseases for the irritation and arsenic existence in the smoke. The immediate effects were staggering to the physician. Hamilton described sulfur dioxide as "the most painful of industrial gases," as the irritation felt like a "red hot stream poured down the throat." Blood sputum usually followed severe exposure.[63]

The reports on the different effects of sulfur and metals in workers' bodies were headaches for mining firms such as the US Company but gained Hamilton international praise. In December 1918, one year after the publication of her report on explosives, she received in Hull House a letter from David Edsall, the dean of Harvard Medical School. Edsall was able to obtain funds for five years of research from the Retail Trade Board, in order to appoint her as the director of the research on the conditions of health in department stores. After additional negotiations with the medical school, the president of the university, and other faculty, they decided to nominate Hamilton as assistant professor of industrial medicine. On March 10, 1919, the president and fellows of Harvard College signed the appointment of Alice Hamilton for three years. She was the first female professor in the history of Harvard.[64] The effect of sulfur and lead on workers' bodies started to be well known in the following years, but the respiratory diseases in the workers' bodies studied by Hamilton surged even more in the next decade, produced by another molecule also present in mining operations: silica.

HOSPITAL MINERO, REAL DEL MONTE

On January 29, 1919, Dean Stanley Calland, general manager of the United States Company's Mexican division, filed the paperwork to make up a new merged mining property, composed of fifty claims in Mineral del Chico. The old mining district had been, in the years before, heavily explored by the US Company, but the problems of the Mexican Revolution and the First World War had delayed most of the firm's expansion plans. In the previous years, the company employed consultants from several US universities—George O. Scarfe, Orrin Peterson, and Fred Searls Jr.— and financed a report from Dr. Carlton D. Hulin and Edward Hollister

Wisser, to predict the direction of the veins between fractures in the northeast and the resources in the Pachuca Valley. As the lead-silver sulfides found in Mineral del Chico and Real del Monte were similar, many engineers in the company believed there were signs of a common fracture between the two historic districts. The new mining consolidation was the first step in expanding the firm's activities, hoping for a common mineralization of the territory between Pachuca and El Chico. The name that Calland chose for the new claim was "La Influenza."[65]

Over the previous year, "the Spanish flu" had decimated Pachuca and Real del Monte's mining population, as the pandemic hit urban centers across the globe. In October 1918, Mexico City, the port of Veracruz, and Laredo in the north reported the arrival of the flu epidemic. The spread followed the routes of the globalized economy in the country. Cosmopolitan centers, linked to the global economy, had suffered, especially in the winter of 1918 and the beginning of 1919. In the mining town of Cedral, San Luis Potosi, the county had to purchase special lands to bury the growing number of deaths by the disease; coal mines in Coahuila and the oil fields in Veracruz were abandoned; authorities in Sinaloa stopped communications by railroad with infected towns, isolating the mining district between Guanacevi and Hidalgo del Parral, in Chihuahua; and authorities in Yucatán had to send doctors to different henequen plantations infected by the new flu strain.[66] The national death toll was around a half million deceased, just in the last three months of 1918.[67]

The mining camps in Pachuca and Real del Monte were among the most affected in the country. The Hospital Civil in Pachuca, located on a US Company property and partially financed by the firm, established a special service for treating the sick.[68] In November, with the deaths of some government officials and teachers, the state government closed schools and churches, implemented a lockdown in the town over gatherings, made compulsory the isolation of the sick inside their homes, instructed the public to wear masks in any interaction in public spaces, and forced prison inmates to dig mass graves for the deceased.[69] Calland reported to company headquarters how the infection affected the firm's clerks and administrators, especially in November and December. However, those most affected in the mining camps were the workers. All the firm's divisions in North America experienced similar outbreaks in the previous months, as the pandemic arrived with full force into the Salt

Lake Valley by the middle of October.[70] Similar measures, like the use of masks, the confinement of the sick, and the control of gatherings, were put in place in Salt Lake City and in Bingham, and the Red Cross installed special flu hospitals in the mining camps for taking care of the sick.[71] Other multinational corporations were experiencing similar outbreaks: the De Beers mine, in South Africa, lost four thousand workers in one month, a third of the entire workforce, to the virus.[72] By the end of 1918, the *Engineering and Mining Journal* estimated that influenza was five times deadlier than the war itself.[73]

When Calland named the new consolidated claim "La Influenza," most cities were just coming back to a new normality. While the mining camps in Bingham and Pachuca reestablished their activities, it was not until March that the epidemic was over. In the next years, the firm continued to report cases of the disease, and they often developed as pneumonia and meningitis. The staggering number of cases, both in the underground and in the smelters, never ceased in coming years. Week by week, managers in Salt Lake City reported to the headquarters on the miners' health, as the disease increased the turnover of the workers who left the camps for fear of infection. Just in the last week of March 1925, twenty-five miners in Bingham had meningitis, and pneumonia often followed lead poisoning cases in Midvale.[74] The infection remained in the air of the mines, or so it appeared in the reports of the managers of the firm. Soon, infection and toxicity merged in both physicians' and social reformists' interests on both sides of the border.[75]

Respiratory problems were not a new phenomenon in mining camps. Paracelsus identified the origin of the miners' pulmonary diseases in metal vapors of elements, not as punishment by sins committed against mountain spirits, as previously believed. The alchemist named chaos the air formed between the earth and heaven, which nourished the lungs of man. The stars ruled this chaos on the surface, which imprinted itself into men's bodies, beginning in their lungs. There was different chaos below the earth, which ruled the lungs of those who breathed in the mines: "Likewise, there are the mineralia of the earth, the heaven and the stars, and they rule their chaos in the same manner as the outer heaven its chaos."[76] This imprint of chaos only implied disease because it destroyed the order between the Macrocosmos and the Microcosmos, the interior and the exterior. For Paracelsus, disease and poison were about order, not about substance. In his peculiar view of creation, God had created the world by organizing matter, *separatio,* but everything was present in everything. Sickness was a rupture of this order, a printing of the outside into the inside: "Man has a skin; it delimits the shape of the human body, and through it, he can distinguish the two worlds from each other—the Great World and the Little World, the macrocosm and man—and can keep separate that which must not mingle."[77] In other words, lung disease was the expression of the mineral body of the mountain into the physical body of the workers.[78]

The influenza pandemic uncovered the long-term impacts of mining in the lungs of the miners. Miner's consumption was a well-known condition for mine workers, but the use of industrial crushers and pneumatic drills had noticeably exacerbated the problem. Lung disease in the workers' bodies drew a continuous line into the mining camps between the eruption of global pandemics and the circulation of industrial technologies. Over the following years, the internal deposition of minerals into the miners' bodies became completely convergent with the detection of minerals in the mountain. The geologists of the company calculated the concentration of silica in the Pachuca ore around 60 percent on average, so most of the matter freed during blasting contained the harmful breathable particles. Below the surface, the safety engineer, L. R. Jenkins, installed konimeters to measure the air's silica contents under several conditions. The measures allowed the company to track concentrations and try to keep it below 300 particles per cubic meter. On the drilling machines, the company had installed dust collectors and implemented wet drilling, reducing the liberation of dust under the pneumatic drilling considerably. They sprinkled the surfaces before and after blasting, and they piled the ores into mud after the first crushing inside the pit. Ventilation was also crucial to helping the particles' dispersion, controlling the distribution of fresh air through the installation of fans at "governing points." Management also tried to control the matter on the surface, as the crushing machines released free particles into the Loreto plant. In the 1920s, E. L. Young had installed a baghouse, under the patent developed by the corporation in Utah, and adapted it to filter silica particles and sulfide dust.[79]

The tracking of the dust in the mountain, the mineral space of the corporate body, continued with the tracing of the particles' deposition into the workers' bodies. At the beginning of their employment, the doctor at the hospital took an X-ray of the applicant's lungs. The negatives were not accessible to the worker, and the company doctors did not regularly examine the exposed miners. By the 1930s, the company had a collection of at least three thousand X-ray plates from workers, detailing their chests' contents. The archived plaques were used only to compare with the lungs of a sick miner, if he subsequently tried to sue the corporation. On occasions, the company tried to measure the detectable particles of silica in workers' bodies up to their death. In 1931, the Mexican government declared silicosis an occupational disease, and if a miner's death was attributable to silicosis in a lawsuit, the body would have to be subject to a necropsy to analyze and weigh the lungs affected. Sanchez Mejorada, the company's attorney, offered authorities the use of the corporate hospital and the assistance of the staff to examine the dead bodies and conduct the measurements of the organs.[80]

In the opinion of Numa Espinola, a Mexican doctor in the company's employment, the routine examination of the lungs was not enough to assess the damage. In 1935 he maintained that the corporate hospital's staff should analyze the dead bodies alongside the authorities, using microscopic techniques. When the silica particles entered the lungs, they created specific damage, and they would be surrounded by different layers of macro-fagus, conforming to a granular appearance around the crystals. The geometry of these nodules will invade, on a microscopic level, the pulmonary tissue, limiting the filling capacity of the lungs and their processing of oxygen. The presence of silica in the lungs or in the sputum was always a sign of the pathology's development by the particles of quartz, as most of the aspired crystals were expelled through coughing. Only a microscopic analysis of the tissue could differentiate silicosis from tuberculosis and establish their relative sequence. The temporality of lung scarring for Numa Espinola, as mineralization for Waldemar Lindgren, was central to assess the effect of mining in the miners' body. Asking for that kind of detail highly constrained the success of lawsuits against the company, as the forensic tests would have to be performed in Mexico City. This level of forensic evaluation was not available to mining camps, and in the country only one study, by Tomas G. Perrin, professor of histopathology at the National University, was able to identify the different types of damage on the tissue of 160 bodies of deceased miners in 1933.[81]

The intervention of the corporate hospital into the workers' body appeared still as a noble touch, a therapeutic exercise of sovereign powers through science and planning. The appearance of occupational diseases in the public sphere implied that the hospital had changed its primary focus from the treatment of trauma to the control of the substances in workers' bodies. Gold, silver, lead, zinc, copper, coal, arsenic, and other minerals appeared in the earth's crust, combined into complex minerals, and into ore bodies that had to be extracted and located. These compounds could also be found in the mines' human subjects, like silicosis in workers' lungs. Matter and subjects of the Leviathan appeared confused in the lungs of dying miners, and the company had to monitor the existence of the minerals from the mountain onto the deathbeds of their workers. The corporation designed a method to distinguish the chaos of these two dimensions, cosmos of the space and the cosmos of the subject: a necro-geology within workers' bodies.

CAMBRIDGE, MASSACHUSETTS

In January 1937, Charles F. Moore, president of the foreign operations department of the United States Company, evaluated several reports on the firm's Mexican mineral bodies and deposits. He reviewed the article "Ventilation of Small Metal Mines and Prospect Openings" by Oscar A. Glaeser, published in volume 126 of the *Transactions of the American Institute of Mining and Metallurgical Engineers*. Glaeser was an expert on ventilation, toxicity, and dust explosion in mines and cyanidation plants in the West, an associate mining engineer at the Bureau of Mines Experimental Station, and a professor at the University of Washington.[82]

The paper had been presented two years before in the AIME conference, and, although not cited, the data came from the firm's operations in Mexico. Moore himself had sent Glaeser information on their efforts to detect and prevent silicosis in the mines in Mexico. E. L. Young, the mining director at the Mexican division, enumerated the diverse mechanisms of detecting dust underground and diagnosing silica in the lungs. In their letter to Glaeser in the summer of 1935, Young and Moore argued that even if the company's name did not appear in the publication, the paper would attract undesired attention from the authorities of Mexico and the United States. Glaeser wrote back assuring that the data would remain confidential, and the author of the paper would be one of his students, Vito A. Brussolo, not him. Nonetheless, the extract referring to ventilation in the mines two years later listed Glaeser, not Brussolo, as the author.[83] The corporation's anxiety about the dissemination of the firm's internal data revealed the relations between the internal structure of knowledge inside the firm and the external creation of science by university professors and government officials. As we have seen earlier in this chapter, company officials regularly shared information with experts from the government and universities, establishing long-standing relationships with the geology and medicine departments in several of them. In exchange for obtaining a corpus of company research, the firm used some of the expertise and, in legal cases, testimony from university and government technicians. Glaeser's paper appeared as a transgression of the collaboration between the expert bodies and company officials to trace silica in the space and the subjects of the corporate sovereign.

The second concern of Moore, during those days in 1937, was to evaluate the mineral bodies below the Pachuca and Real de Monte properties. There had been internal discussion over the future developments of the division and the direction of further exploration. In February, he contacted Warren Judson Mead, a structural geology professor at MIT, to write a geological history of the mining district. The contract was for an indefinite length, at a rate of one hundred dollars per day, plus transportation and living expenses. Mead arrived in Pachuca in May with his assistant Olaf N. Rove, and was received by the company's officials, M. H. Kuryla and E. L. Young, as well as C. L. Thornberg, the staff geologist.[84]

In the following months, Mead surveyed the district with airplanes and stereoscopes, allowing him to reconstitute the region's mosaic geologic map. The landscape and altitude of the mountain allowed him to rebuild the history of the rocks, but part of his new analysis methods required the microscopic observation of the rocks, following patterns at the microlevel and relating them to the macrostructure. Signs were not only on the surface of mountains and valleys but also on their internal and invisible composition. To be able to access this hidden history, he had access to an unusual archive of the mountain, a lithotec containing rock samples from different deep locations in the range gathered in the mines over the previous decades. He spent more than three months in the area with the staff's help and came back to Boston at the beginning of September, in time for the beginning of classes.

He was used to combining his academic and administrative responsibilities with consulting work for corporations in the United States. Like some other colleagues at MIT, he had consulted for the Aluminum Company of America since 1912, but he also had experience in colonial projects in the tropics and the East. He was part of the commission that experimented with Panama's Canal after some slides forced the project's closure in 1915.[85] Six years later, Mead took part in evaluating the geology of the coal district of Fushun, part of Japan's rights in Chinese territory after the war with Russia. As had Richard Strong, Mead had worked under the authority of the South Manchuria Railway Company. He combined the district's magnetic and visual examination with microphotographs of thin sections of the different ores, locating promising areas and ores, and possible ways to process them. His work on the areas of Ta-Ku-Shan and Oten were vital in the iron bonanza of the Anshan Works, which became a growing center of attention for the Japanese Empire until the military occupation in the 1930s.[86]

Empires, nation-states, and corporations were just different scales of planning for Mead and other technical experts, but they remained relevant in the sovereignty spheres of the era. Structural geologists were also crucial in designing the hydraulic projects of the New Deal. Between 1932 and 1937, Mead took part in planning the Federal Emergency Administration of Public Works projects on conservation and flood administration, visited thirty-seven dam sites, and consulted for the US Army Corps of Engineers. That year he took part in the maintenance of those projects and supervised the works on the dams built in the Panama Canal around Madden Lake.[87]

"The major problem is the working out of an acceptable history of the earth movements which have fractured the rocks of the region," he stated, discussing the work of previous experts. All around Pachuca and Real del Monte, the geologist found traces of the earth's violent movements, such as mineralization and fissures. A Cretaceous lake had caused sedimented rocks on level ground territory, but volcanic and tectonic activity had folded the crust and created faults. Successive volcanic eruptions distributed new rocks, and hydrothermal phenomena cracked the crust, mineralizing with metals in the veins. The mountain, valley, and the order of rocks revealed the marks of the magma movement. The macro- and microgeometry of the different ores involved, the separation between layers of rock, and the shape of the horizons revealed events over eons.[88]

This succession of geological events was fundamental in determining the mineral structure of the area covered by the Mead report.[89] The order of the past revealed the structure of the territory, and that structure shaped the form of the future. Other geologists hired by the firm believed that the district's metallic veins were uniform, as the faulting occurred before mineralization, and mineralization was, itself, a briefer event. This implied a correlation between following the faults in the district and the minerals found in them. The post-mineral faulting theory, proposed by Mead, implied no correlation between the faults, and that the ores found around the district were heterogeneous. Mead's suggestions were to abandon further exploration to the north of Pachuca and Real del Monte, as the minerals located there, particularly dacite, were extruded after the mineralization, implying that the fractures there were not filled with metals.[90] The guide for doing underground exploration in the future was to evaluate the mountains' external characters: avoid the gray colors of the dacite and do further exploration in the red and darker stones along the surface. After Mead's report, company officials decided to increase the exploration work on the northwestern parts of the area, close to Mineral del Chico, instead of exploring the Real del Monte area around the Arevalo mine. In the coming years, the company increased its reserves in the northwest, as advised, and integrated the sector into the underground centralization project around Real del Monte.

After the report, the firm's officers updated their formal relations with the geology department of MIT. As head of the geology department, Mead continued to be one of the crucial links between mining corporations and the institute. In 1941 he convinced the US Company to give three research fellowships annually to MIT focused on developing spectroscopic investigations with crystals. By then, most of the major corporations—ASARCO, Kennecott, Phelps Dodge, the US Company—had fellowship programs at MIT for students in geology and metallurgy, and made annual contributions for research. Again, the war had set the university into a privileged position between the industrial and wartime uses of technology. For example, Mead analyzed for the mining companies the potential uses of radar technology into surveying below the ground. In 1948, the Committee on Business Corporations at MIT transformed into the Industrial Liaison Office, seeking grants-in-aid from industrial corporations. Under the new Industrial Liaison Program, corporations made contributions to support research at MIT, and, as a quid pro quo, the university organized conferences for managers and technicians at the firm. Members of the liaison program staff visited the companies' operations, gave suggestions, and discussed the research priorities with the firms' research and development teams. In addition to the firms' usual consulting jobs, researchers would eventually reach out to corporations to get funding for new research projects, and the companies could keep the patents. In 1949, the US Company joined the Industrial Liaison Program.[91]

During those years, Cambridge's elite schools of higher education had established formal mechanisms for firms' financial contributions with a reciprocal relationship with students and faculty. In 1924, Richard P. Strong and George C. Shattuck, from the Department of Tropical Medicine at Harvard, organized a subscription program for firms that contributed a minimum fee of $2,000 per year to the program. In their friendly reminders to their corporate sponsors, Strong would "hope that if at any time we can be of service to your company in connection with problems relating to tropical or exotic diseases, that you will not hesitate to call upon us."[92] At the end of the 1920s, the department shifted its interest from mining operations to rubber companies and consulted on the creation of the Henry Ford Hospital in Fordlandia, Brazil, and the exploration of diseases in Liberia for the Firestone Company. When asking for the reimbursement of the African Harvard Expedition expenses from the corporation, Strong attached a copy of the compilation of studies and a list of the 140 libraries around the globe that received the book. The publication was dedicated "to Harvey S. Firestone, who has done so much for the development of the country of Liberia and for the Welfare of its inhabitants."[93]

Corporate officials also increased their alliances with industrial physicians in elite schools by offering the same benefits that had their geology, mining, and tropical medicine counterparts. In 1935 Alice Hamilton retired from her position at Harvard, and in the coming years, her department changed alongside the Harvard Medical School. As with the tropical medicine department, led by Strong, the industrial hygiene department strengthened its relations with firms, industry lobbies, and military officials to design protective mechanisms against germs and the environment. Joseph C. Aub took Hamilton's place as the main expert on industrial diseases and established a stable relationship with the Lead Industries Association. In 1946, the Harvard School of Public Health became independent from the Harvard-MIT School of Health Officers and absorbed the Departments of Tropical Medicine and Industrial Hygiene. That year, Aub organized a symposium on lead toxicity and was one of the main skeptics of the harmful effects of lead in pregnant women and children, recommending the use of white lead on babies' cribs.[94] "In your relationship with industry, the reputation of operative discretion, which inspires confidence, is pretty essential," he wrote in a letter to Harriet Louise Hardy, one of Hamilton's pupils and the first tenured female professor at Harvard Medical School, following a dispute over an opinion with her over cancer research and occupational medicine. "If industry gets the impression that their problems will be widely discussed, they will become uncooperative," he noted.[95]

By then, industrial medicine was a mainstream concern for any industry that used metallic sulfides in their processes, and firms established pipelines between academics working on industrial diseases, government agencies, and industrial relations departments. Mine safety experts, doctors, and geologists mediated the corporate body's relations with the mineral deposits from the mountains to the workers' lungs, and scientists incorporated professional consulting in their research on an institutional level. Safety engineers, physicians, and geologists tracked the transmission of materials into the different bodies of the corporate sovereign.

Also in 1946, the United States Company elected Oscar A. Glaeser as part of the board of directors of the firm. After the disagreement with Moore, Glaeser was able to reconstruct his relationship with upper management in the Boston headquarters. He worked as relations manager of the western operations of the US Company for some time, and would later become general manager of the western division.[96] The professional career of the safety engineer had been built around a good relationship with firm management, even after the controversy with Moore, and he was able to reserve, in the next years, delicate information on hazardous gases and particles in the lungs derived from his investigations. His transformation into a board member, at the head of the corporate sovereign, was only natural. In 1950, Glaeser published again on mineral deposits in workers' bodies in the *Industrial Hygiene and Occupational Medicine* journal. He avoided negative terms, such as "hazardous" or "explosions," that appeared in his earlier works. This time, the paper naturalized the changes in miners' bodies as they contacted the chaos in the underground, the silica dust in sulfide ores deposits, and how that shaped their relationship with the corporate body. The title read, "Compensation for lung changes due to the inhalation of silica dust."[97]

The work of Glaeser, as safety engineer and later industrial relations manager of the firm, was both an old and a new profession. On the one hand, it was as ancient as medicine and alchemy themselves, figuring out the problem of the transit between the mineral deposits from the depth of the shafts into the tissues scarring in the workers, tracking the transfers of sulfur and silica from mineral bodies to human bodies. On the other hand, it was a new phenomenon inside the corporate sovereign. After the Second World War, more technicians had to become experts in negotiating damage of worker's bodies with the political bodies organized by unions. The authorities at MIT, for instance, occasionally proposed arbitrators for labor negotiations with the firms in their Industrial Liaison Program.[98]

They shared more than economic interests, social status, and professional spaces with the technical officials at the US Company, as they all viewed themselves as self-regulating bodies of knowledge, technical planners above the conflicts of the political sphere. Some would later become part of the company's permanent staff, while most would remain in the consultancy business. Neither their links with other sovereign entities, such as empires and nation-states, nor their employment in elite schools, inhibited their transits into the corporate Leviathan. Their diagnosis and prognosis predicted the artificial animal's fate by examining the contents of its different bodies: the ore bodies in the earth as objects, the political bodies of workers as subjects, and the metabolism of minerals in the workers' bodies. Their knowledge was vital to expanding the organization into the world, tracing matter and subjects inside the corporation, and forming the body of sovereigns in the corporation: the technical officials.

CHAPTER FOUR

Soviet of Technicians

> So still he seems to dwell nowhere at all; so empty no one can seek him
> out. The enlightened ruler reposes in nonaction above, and below his
> ministers tremble with fear.
>
> —Han Fei, *The Way of the Ruler*, ca. 200 B.C.

BOSTON, MASSACHUSETTS

On Monday, March 3, 1919, Sidney J. Jennings attended a presentation at Boston's section of the American Institute of Mining and Metallurgical Engineers (AIME). One month before he had been named president of the institute, the largest engineering association in the country and with influence abroad. Later that evening, Jennings had dinner with the panelist, F. W. Draper. They both studied engineering in Boston, Jennings at Harvard and Draper at MIT, and shared experiences south of the border. Jennings had been the vice president of the exploration department of the US Company, and regularly evaluated properties in Mexico, Peru, and Honduras. Draper, by his side, lived in Mexico between 1902 and 1912, working in several metallic firms in Durango, Coahuila, and Zacatecas, and in 1911 he tried his luck on a mining operation on his own in Jalisco, Mexico. That day, however, Draper was not scheduled to talk about his experience south of the border. In 1912, after a year of trying to make the firm profitable during the armed conflict of the Mexican Revolution, he had moved permanently to the Ural Mountains, in Soviet Russia.[1]

The eastern country was in everybody's mind in Cambridge those years, following the Bolshevik Revolution and the founding of a new Communist International. The emergence of a centrally planned economy intrigued the intellectual circles in the United States, especially technicians who both feared and admired the dramatic changes of the era. That year Thorstein Veblen published in *The Dial* a series of articles about the role of engineers in modern society. In his last piece, "A

106

Soviet of Technicians," he imagined the possibility of a socialist revolution in the United States. From Veblen's perspective, in North America, the movement could not be a worker's rebellion, but a conjoint action of the technicians against the absentee owner class. The idea of a "Soviet of technicians" derived from Veblen's critique of the leisure class and was informed by his experience in public administration during the First World War.[2]

Technicians had experienced a fair share of political power in the first decades of the century. In his last years in Mexico, Draper had started to provide occasional technical advice for the Verk Isotz Estates near Ekaterinburg, Russia, regarding the exploiting of platinum mines. The famous mining engineer and speculator, John Hays Hammond, tried to organize an amalgamated platinum company in the Russian Empire in the last years of the nineteenth century, inaugurating a flow of US metallurgists into the East.[3] In the next years, the small US community in nearby St. Petersburg grew with the arrival of US firms into Imperial Russia, and they built an American corporate bubble in the city.

Like Veblen, during the war, Draper experienced the capacity of planning by engineers. Russia produced up to 95 percent of all platinum worldwide, and the metal was essential during the war in the production of sulfuric and nitric acids, which were used in the manufacture of explosives. The strategic nature of metals during the war combined state sovereignty with planning by mining traders. At the beginning of 1917, as the future involvement of the United States in the war became increasingly apparent, German agents tried to secure as much metal as they could in the global market. As events of the revolution and war unfolded, Draper endeavored to accumulate platinum in St. Petersburg through direct relations outside the market, and to direct the delivery in the most challenging way: by shipping it to the East. With the support of the US embassy, Draper conducted the shipment through Siberia, Manchuria, and the coasts of Japan, changing passports after the Trans-Siberian Railroad. As in Mexico, Draper was escaping a socialist revolution. Soon after the convoluted departure of Draper from Russia, the Bolsheviks took control of all technical installations and staged a coup at the Winter Palace in St. Petersburg.[4]

Jennings and the Boston section of the AIME could relate to the deployment abroad and the direct control over entire sectors of the economy. The experience of Draper in Mexico and the United States shaped his answer to the conflict in Russia. During the First World War, the traffic departments of mining firms negotiated directly with the US government to distribute the shipment of products. In contrast, mining firms in Mexico had to ship and take care of the railroad routes themselves between 1911 and 1914, as railways were a frequent target for the different factions of Mexican revolutionaries. Mining engineers working in Mexico during the revolution often took direct control over railroads to ship metals for the war, and they also led the reconstruction of the communication system between productive units and the US market. In 1922, Anaconda, ASARCO, and US Company departments pushed the Southern Pacific Railway Company to rehabilitate their freight lines in the US West and the Mexican state of Sonora. After the joint negotiation, the companies not only improved the lines affected by the armed conflict but also connected Arizona with Hermosillo through Nogales.[5]

The organization of the corporation—its form—was never an abstract set of rules and functions: it was always a concrete historical actor. The increase in the power of mining corporations in the multinational scene was led by their technical officers, and they designed a political body of decisions in the interior of the firm, replicating the structure of the holding company and its divisions. In the case of the US Company, the division managers in the 1920s and 1930s took part in the vice president's board, which decided the general strategy of all the divisions. The general officers of the holding company served as a coordination bridge between the technical officers in the subsidiaries. The traditional scheme described by Alfred Chandler, with its pyramid of authority and hierarchy, existed only at the beginning of the corporations, immediately after the first mergers. In contrast, by the 1910s and 1920s, officers operated more as consultants than authorities concerning the different departments of divisions. Geologists, chemists, engineers, and agronomists of each unit exchanged information within and between divisions, as well as examining new properties, innovation in processes, and changes in the organization of the productive units.[6]

The officials at the divisions bargained for access to capital, new technicians, and supplies with the holding, and created social bonds and circulation of knowledge, besides the hierarchical relations. In short, the manager's system within the US Company was at the same time a technical and a social system, a political and economic world, a closed and international space. This complex tissue of vertical and horizontal relations, in an expanding but also restricted organization, had as a premise the standardization of the different components. Transfers, control, and training were the ways the multinational corporation homologated their personnel into a global structure: a sovereign body regulated by technicians.

This chapter traces the continuities between the expansion of American mining engineers to Mexico and Russia by multinational corporations, using the case of the United States Company, and the operation of centrally planned societies. In the first quarter of the twentieth century, multinational mining corporations established circuits of technicians among their divisions, designed training plans of technical crews, built communities of technicians abroad, promoted the exchange of experience among their engineers, and developed a consciousness of their roles as technical and economic planners within a Leviathan. The final design of the US corporation in the first decades of the twentieth century was not that far from Veblen's dream of a centrally planned society ruled by technicians. For most firms, shareholders had no direct intervention in the decisions of management; hence, engineers, managers, and technicians in general had full power over the scale, scope, strategy, and design of the firms. The first experiment of this integration was not only national, inside the United States, but also regional, particularly into Mexico. By the second decade of the twentieth century, this "soviet" of North American engineers expanded, as experienced by Draper, into Russia even after the Bolsheviks took the Winter Palace and created a new kind of Leviathan. Similar to the final trip of Draper through Siberia, the construction of this international body of technicians required the material movement of people, goods, and technology through the railroad network.

CIUDAD JUÁREZ, CHIHUAHUA

In June 1935, Kein W. Emerson entered Mexico through the border with El Paso. He held a fake passport provided by the agent of the traffic department of the US Company. In a series of maneuvers that echoed Draper's experience twenty years earlier, after crossing the border, Emerson changed his passport and contacted Carlos Sánchez de Mejorada in

Mexico City, the general attorney of the Mexican division of the US Company. He announced his arrival to the managers of the division in Pachuca and took the train with the ticket provided by the agent in Ciudad Juárez. In the previous months, the general manager of the subsidiary and the vice president of the Mexican division of the US Company, M. H. Kuryla and C. F. Moore, respectively, had negotiated the transfer of the American engineer from Boston to Mexico. They consulted the vice director of the international division, S. B. Douglas, letting him know that they planned to offer Emerson a salary 33 percent higher than their original offer. Douglas counseled Kuryla to double the increased offer. "I see no purpose in having him go to Pachuca unless he feels that by so doing, he is being promoted and that the salary is entirely satisfactory," he argued.[7] After a raise in the offer, Emerson accepted the job. The transit with false documents was the standard practice of migration inside the US Company. This transgression of the territorial sovereignty allowed the technical crew of the firm to start the work while the legal team finished the paperwork. Three months later, Emerson's family joined him, fully regularized.

The border-crossing episode reveals several traits in the communications planning that multinational firms operated in North America. In previous decades, most of the multinational mining corporations had invested in the construction of significant communication mechanisms across borders, and many technicians had to personally take care of shipments of concentrates. The initial growth of the big mining multinational corporations in the Americas would not have been possible without the development of continental, cross-border railroad systems. The arrival of the railway to Utah unchained the first rush into Bingham Canyon in the 1870s, and the integration of the Mexican railroad to the United States in the 1890s fueled the investments of big American corporations in north and central Mexico. Nonetheless, the lines established by railroad companies were rarely sufficient to support the plans of mining corporations in the territories. Mining firms invested directly in the construction of some lines, when it was impossible to push the state and other firms to do so. In 1891, the Anaconda Copper Mining Company opened the Butte, Anaconda & Pacific Railway for better corporate communication in Montana. Two years later, the new railway company opened its first lines, not only providing freight to metallic mines but also transporting passengers and coal. The company grew in Montana during the subsequent years, controlling all the lines to Anaconda and extending to the connections to Colorado and Seattle to create an integrated transportation network under the company's control.[8]

In 1907, the Alaska Syndicate started the construction of a line from Kennecott to Cordova, for 196 miles, transporting concentrates treated in a small plant in the glacial area. Two subsidiaries of the Kennecott Corporation made the shipment to the continental United States possible. The Copper River and Northwestern Railway Company built and operated the railroad to interior Alaska, and the Alaska Syndicate merged the Alaska Steamship Company, the Northwestern Steamship Company, and the Northwestern Fisheries companies, transporting the shipments from Alaskan ports to Seattle. The Kennecott Corporation, in short, controlled the mineral production in the territory and most of the transportation routes to Alaska in those years. The railway was the base for the operation of the mines and the technical model for the railroad to Fairbanks. The rest of the mining companies depended on the Alaska Steamship Company, even after the Kennecott mine closed in 1938.[9] The same strategy played out elsewhere in North and South America. Far to the south, in Chile, the mines of El Teniente and Chuquicamata required the construction of a large railroad connecting the isolated desert of Chile with global markets. The Anaconda Company sent already processed copper to the port of Antofagasta, a distance of 165 miles. At the beginning of the operations, the corporation purchased the Chile Steamship Company but transitioned in the following decades to establish their own steamship lines.[10]

The US Company is another example of investment and control over shipment routes. In 1913, they acquired the Consolidated Coal Company, which had an infrastructure of railroads ranging from Carbon County to Provo, Utah. In the coming years, the US Company divided the firm into two subsidiaries, the US Fuel Company and Utah Railway, which operated independently. While US Fuel remained a rather terrible investment for the corporation, the Utah Railway freight charges were stable, and it could transport other merchandise in the 1920s. The Utah Railway improved the relations of the US corporation with the Denver and the Rio Grande Company, establishing long alliances for the renovation of their standard lines.[11]

Nevertheless, after the trial of the Standard Oil Company in 1911, mining corporations restricted their participation in railroads to small lines and intervened in the general market by establishing traffic departments. These departments concentrated the information about rates in the United States and South America, for different freight and distances. They would then purchase and pay for the shipping, track the merchandise, take care of delays, and build agreements with communication companies. The traffic department of the US Company had agents widespread in the territory, from Seattle to Laredo, and from New York to Valparaiso. It controlled not only the metal shipping from the smelters to the markets but also coordinated the internal traffic between widespread units of the firm. The officials of the department planned the movement of concentrates to plants and refineries, the shipment of supplies and technology, and the payment of taxes for imports and exports. Moreover, the officers audited the movements of the technical staff of the company and arranged their migration status when crossing international borders. These functions required high specialization and technical expertise. The traffic department of the US Company alone gathered information about six hundred different tariffs for transporting minerals, machinery, people, and even letters and personal packages.[12]

As necessary for the homogenization of operations among the divisions was the consolidation of purchases through the stores departments. In 1904, the US Company acquired the Bingham Mercantile Company, a small store in Bingham Canyon in Utah, to provide some mining equipment for the local mines and consumer goods for workers. Soon, the firm transformed this small unit into the United States Stores Company, which was an intermediary for all its corporate divisions for handling the purchase of materials, profiting from centralized knowledge of prices in different industrial cities in the United States, and of transportation fees and time requirements. Combined with the traffic department of the parent company, the US Stores had agents all over the United States, Alaska, and Mexico. The consolidation of purchases through this department reduced prices and allowed the parent company to control the expenses of the divisions and, at the same time, unified the subsidiaries technologically.

The map shows the complex internal circulations of products made by the United States Stores in the 1920s and 1930s. In terms of value,

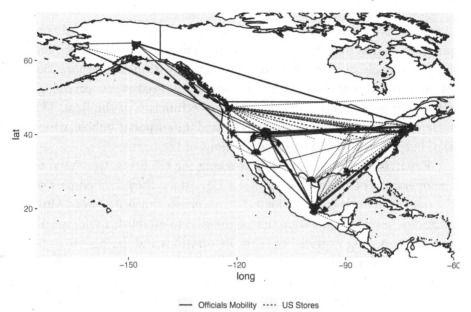

— Officials Mobility ···· US Stores

Multinational transits in the US Company. Data from AHCRMyP, sección compras, almacén, facturas de compras, vol. 2; UAF, USSRMCo Records, 1924–61, Purchase Orders, boxes 1–6; USSRMCo, *Annual Report*, 1908–50.

the most important component of the shipments overseen by the US Stores were machinery, spare parts, and specialized materials from manufacturers in the industrialized East and Midwest. The operation of the US Stores surpassed, as with the transit department, the strict logic of vertical integration, as it also centralized and standardized the consuming practices among engineers. Alongside gold dredges, cranes, and seven-thousand-gallon pumps, company officials turned to the United States Stores Company for the shipping of needles and thread, lady's wristwatches, radios, tennis rackets, electric heaters, motorcycle supplies, cans of Paraguay tea, and furniture. This division was no exception within the major mining titans of the era: Anaconda Copper, through the Mountain Trading Company, and Phelps Dodge, through the Phelps-Dodge Mercantile Store, followed similar practices.[13] The construction and maintenance of a corporate bubble was, in this way, both a company policy and a self-designed blueprint by the company officials.

To be sure, the productive features of the corporation were a composition of industrial mines, big and advanced plants, and efficient railroad systems. The transit and stores departments illustrate the internal communication and the network economies of the multinational corporations at that time, the scale in the transmission of information, and the critical role of planning and coordination within the corporation. Mineral and capital movements were just as essential as the technological exchanges that operated through the technicians of the firm. The maintenance of those exchanges thus created the corporate bubble, which was, in time, reinforced by some divisions of the firm.

Expatriates such as Emerson, working for US firms, were vital to the operation of the corporation as a Leviathan: they formed part of the operation both as technological components and as its rulers. Multinational mining firms were one of the first to establish a widespread system of moving experts between the divisions and the parent company in the twentieth century.[14] The line between management and the technical crew was nonexistent, as most of the top managers of the firm were engineers from the ranks of the firm. Usually, top staffers of the parent company held several obligations in distinct divisions, obligating them to maintain coherence among regional subsidiaries. Finally, long-term employees usually followed mobile paths in their careers, circulating between the several divisions of the company. The corporation was, in this way, not only a hierarchical structure but also one that facilitated horizontal relations between the technicians and managers. Instead of recurring to the market, the subsidiaries would bargain with the holding company the transfer of highly trained staff, competing internally with other divisions of the firm. The result of this version of corporate politics, not an internal technical market, decided the allocation of promising officials inside the US Company, between competing departments.

The migration of technicians inside the US Company provides an image of the western expansion of mining corporations. At the beginning of the twentieth century, the company engineers were usually educated on the East Coast, hired by the headquarters in New York and Boston, and then migrated to the West for promised positions. Gradually, western-trained technicians, from Berkeley and Colorado, entered the global career system of the company, expanding to Mexico and Alaska. Finally, those who were transferred to Alaska and Mexico had an implicit or explicit promise of return into the central structure of the company. In other words, the expansion of the corporate boundaries to the north and south followed the expansion of the technical frontier. An analysis of the movements of the top officials reveals the highly centralized and multinational system of global officials in the US Company during the forty years of its international operation. The map also shows transfers among organizations and spaces of the firm. The primary circuit connected the Mexican division, the western operations, and the eastern headquarters. The map also represents a strong interaction between the small companies in the West, even more robust than the relations between Alaska and Mexico. In short, the map illustrates an international corporate structure of mobile engineers with two centers: Boston and Salt Lake City.

Inside the corporate Leviathan, the trade-off between promotions and geographical mobility created a bond of sacrifice-reward among the officials in the local enterprises.[15] Transfers could also prevent excessive integration into local societies and the loss of internally formed human resources to other companies. The circulation of technicians was as crucial for the correct operation of the firm as were the fluxes of capital, technology, and supplies. These transits were, in general, controlled by the same departments of the multinationals. But the construction of a homogenized technical and managerial crew inside the firm was different from the effort to build a homogeneous body of workers described in chapter 2. Managers and technicians themselves designed these transfers, movement, promotion, control, and reward methods. It was, in other words, a body designing itself.

The materiality of the reproduction of this technical elite had, naturally, a counterpart in the formation of a common identity as rulers of the corporate body. Edmund Burke described the British East India Company as "a kingdom of magistrates" where the *corporate spirit* predominated over the sovereignties they acted upon.[16] Likewise, the US Company officials were building this sovereign power, a *soviet of engineers* or a *confederacy of officers,* in their design desks and not by violent revolution. This confederacy had not a single space of action but a multinational one, and, as with any sovereign space, they had to consciously build it.

HACIENDA DE CHIMALOAPAN, HIDALGO

It was Fourth of July 1931, and the Hacienda de Chimaloapan was full of young and adult US citizens. As it did every year, the US Company closed its general offices, giving the day off to all Anglo employees and organizing a celebration for all of them. During the day, they organized small races between the children of officials, while bachelors and married officials competed in a baseball match. At night, they organized a gala, dancing with American, Mexican, and British flags. Surrounded by the smoke of the barbecue, the band of the Hidalgo State played the United States national anthem and other patriotic American songs. The Hacienda de Chimaloapan, an old agricultural property of the Conde de Regla, was the outdoors center of the social life of the firm's Anglo employees, their favorite place for barbecues and celebrations. On Thanksgiving, the Anglo employees enjoyed the day off at the Hacienda in the morning. Later, the local Protestant church gave a service, and the top officials of the corporation had dinner at the general manager's house. It was there, as well, that the corporation organized American and Mexican celebrations, constructing its own community but with relations with the local authorities. On September 15, Mexican Independence Day, the corporation provided fireworks and signs with the legend "Viva Mexico" and invited local politicians to the corporate celebration. In return, when top government officials arrived at the city, the company managers were part of the cohort and were usually guests at the state palace.[17]

As a result of the transfer policy of the firm, the lives of the engineers were, at the same time, well established and based in the area and very similar to their experiences in the United States. This was a small colony of a multinational corporation, not atypical of such communities in other locations in Latin America, Africa, Asia, and elsewhere. In effect, when officials of the US Company arrived in Mexico in 1906, they did not find an environment alien to foreigners. Instead, a mountain of employment applications from US citizens, British citizens, and English-speaking Mexicans arrived at their offices, applying to positions as secretaries, clerks, accountants, timekeepers, or assayers, counting on their experience in other mining, commercial and industrial US corporations in the country.[18]

Moreover, the Compañía de Real del Monte y Pachuca had been at the center of the tradition of Anglo management. When the British company purchased the traditional mine from the Conde de Regla in 1824, they sent Cornish engineers and managers to rehabilitate the silver operation, and they stayed even after the British investors sold the firm to Mexican capitalists. The English community reproduced in the next years, mixing migration and integration into business with local families, a traditional path for foreign entrepreneurs in Mexico since colonial times.[19] Their identity, nonetheless, stayed long after the new Anglo community arrived in the last years of the nineteenth and the first of the twentieth century. This time, US citizens outnumbered the English, and while some of them entered the dynamics of integration by marriage into local families, they were eager to construct their own social world. They created separate institutions, sharing their Protestant faith, taking part in secret societies, and defining their yanqui status. The American colony was more self-sufficient than other foreign business communities in the past. Americans were "integral outsiders" in Porfirian Mexico.[20]

In the first years of the twentieth century, the US population in Mexico grew from one thousand to ten thousand, becoming the most significant foreign minority in Mexico, only behind the Spanish. This segregated growth created hospitals, schools, and cemeteries, and a large variety of clubs: the American Club, Society of the American Colony, Women's Club, Young Men's Christian Association, and the Mexico City Country Club. A vital feature in this latest wave of migration was that many of the new arrivals were company officials of multinational corporations, such as F. W. Draper, Sidney Jennings, and Kein Emerson. The new structure of North American capitalism integrated big corporations across the continent and established a continuous flow of technicians from the East to the West, North, and South—and back again. Alumni and professors of Harvard published a *Handbook of Travel,* with articles and advice to help relocated technicians navigate a series of skills from riding camels to identifying hard rocks, from snowshoe travel to anthropological examination of locals, from meteorological observations to the most common diseases in the tropics.[21] The managers and technicians of the US Company, educated in the Boston area, were part of this wave of temporal expatriates.

Commenting on his expatriate life in Mexico, one engineer of the US Company wrote, "There is a fine bunch of 'foreigners' here, and in neighboring camps—much the same type found in U.S. mining camps—and all are ready for reasonable diversion at the end of any week in the year."[22] These expat communities crossed corporate lines, as officials shared practices with other technicians in similar companies. Inside this enclave community, some families and longtime friendships formed, but the loyalty to the multinational corporation surpassed even the integral interaction of these outsiders. Top officials rarely changed employment between the multinational firms in Mexico, and they instead came back to the headquarters.[23] Corporations controlled a more significant part of the lives of their officials than of their blue-collar workers. Top-ranked officials lived in houses of the company, with areas anywhere between one-half and two acres, and sometimes they could accommodate their extended family in the camp.[24]

The teachers at the Elementary American School in Pachuca had paid vacations to attend summer camps to learn Spanish in the summer school sessions of the National University in Mexico City.[25] The older sons and daughters of the technicians could go to the English-speaking schools in Mexico City, with tuition paid for by the corporation, or arrange their enrollment into US schools. The wives of company officials had enormous yards where they cultivated corn, kale, rhubarb, lettuce, cucumbers, onions, and other vegetables, besides some farm animals and chickens. These women created a circuit of consumption and leisure inside the American enclave, driving to Mexico City, shopping, and playing golf regularly in the local clubs.[26]

The structure of corporate expatriation, with a closed community highly dependent on the company for housing, consumption, and cultural practices, was not only a tropical practice[27] but also a convergence with the general model of the movement of managers in the Americas. In Alaska, the company officials were as foreign as they could be among rushers, natives, Russians, and the unbeatable Alaskan nature. The gold placer mines in Nome and Fairbanks could only be worked six months per year and had to restart after the cold winter. The officials who stayed during winter had a lot of free, dark, and cold time during those days. Communications were difficult, as the last boat to Nome, Alaska, had to leave Seattle in early October, and the railroad rarely made trips into Fairbanks after November. By the 1930s, the airplane could still reach both cities for another month but had to stop between December and February.[28] During their stay, however, the company officials took part in the All Alaska Sweepstake, the longest of the local races with dogs and reindeer on over three hundred miles from Nome to Golovin. Officials gathered for the holidays in the Kennel Club, and in spring, they organized the Northwestern Alaska Fair for three days, where they displayed some products of the firm. Often games took place between the white staff of multinationals and the natives coming from the Yukon River to sell their products. In Fairbanks, officials organized panning contests, the traditional placer mining method, between old rushers and new company employees. On Labor Day, the managers made a barbecue in Circle Hot Springs, about 150 miles from the city.[29]

As in Mexico and Alaska, the company integrated culturally with the local celebrations in their metallic divisions in the US West. For Labor Day in Utah, the officials went into the Salt Lake, having a company camp-out for the weekend. On Pioneer Day, the corporation organized the Pioneer and Covered Wagon Days in Carbon County, Utah, representing the history of pioneers, including camping and relationships with indigenous populations, that ended with a feast. On December 22, the officials had an institutional party outside the laboratories at the Midvale plant, and a surprise exchange of gifts. On December 28, in coordination with the neighboring Ohio Copper Company, the Bingham Prospect Company, and Bow Action, the managers and engineers of the firm offered a free dance, dinner, and gathering in Bingham.[30] In the company's division in Eureka, Nevada, the company backed the Club House, where the staffers organized dances and smokers, male-only gatherings with cigars, lunch, and boxing contests. On New Year's Eve, the club organized a dance with the Sunnyside Orchestra, and on Christmas, the school kids performed operettas.[31] The United States Fuel Company in Carbon County, Utah, was present in most social aspects of the life of their managers and their cultural practices. The Carbon County corporation financed three secret societies: the Harrison No. 31 and the Carbon No. 16 of the Knights of Pythias, and the Order of the Eastern Star, chapter Naomi. The carbon division also funded the construction of a local of the American Legion and Ladies Auxiliary, along with the local elementary and junior high school, and coordinated the parent-teacher association. The top officials gathered in the Dinner Club, which only had twenty-two members, and the Hiawatha Gun Club, which performed rifle practices and shooting contests.[32]

The corporation as an entity organized the daily activities and rituals of the technicians, from religion and education to fishing and gardening. The corporation financed local churches in all the divisions; for example, in Copperton, Utah, and in Pachuca, Hidalgo, they financed the Episcopal church; in Carbon County, they built the local Protestant church, and in the Salt Lake Valley the officials coordinated the Old Folks Day of the Church of Jesus Christ of Latter-day Saints. In September, in the divisions of Hidalgo, Alaska, and four localities in Utah—Midvale, Copperton, Hiawatha, and Heiner—the American staff of the US Company organized gardening competitions, with flowers, berries, and squash. The judges were the local authorities and the agricultural experts of the corporation. The officials of the United States Fuel Company and the railroad subsidiary, Utah Railway Company, competed in fishing derbies in the Huntington River. In the Salt Lake Valley, the employees went to the Mirror Lake to the events promoted by the Bureau of Fisheries. In Alaska, the fishing season was at the end of summer, when the mosquitoes were less abundant along the Yukon, Nome, and Snake Rivers. In the fall in Utah and Alaska, managers and engineers hunted deer, and the company provided days off during the best hunting seasons.[33] The US Stores sent books about American sports to all the subsidiaries, along with baseball equipment. The divisions financed golf memberships for their top managers and in places where there were no golf clubs, like Pachuca, Sunnyside, and Carbon County, invested in the installation of them. The company controlled the golf club in Pachuca completely, organizing every two weeks a Kuryla Trophy, named after the general manager M. H. Kuryla. Additionally, the company paid for memberships to golf clubs in Mexico City, where the corporate team usually went into the finals in national championships.[34]

In short, the company managed the most substantial part of the lives of its officials: income, education, mobility, sports, housing, family arrangements, churches, holidays, and consumption. Most of the top managers of the firm were professional mining engineers who had climbed the ladder of corporate authority. As a result, the legitimacy of this regime of technicians was not only mediated for the economic benefits but also by the consciousness of them being the company. As a self-regulatory body, the engineers, managers, doctors, chemists, geologists, and metallurgists created the boundaries of their world by deciding transfers, wages, and benefits, by constructing company houses and establishing interindustry networks. They were integral outsiders in the larger world, but their bubble, to maintain its unity, required the exclusion of blue-collar workers and Mexicans from top positions, the rotation of staffers, and the homologation of their consumption and leisure habits. They controlled the extraction, processing, and distribution of matter, and planned the reproduction of their experience as managers, inside the corporate body. Through that process, they transformed the spaces of the minerals, the bodies of the subjects, and produced themselves as a political actor.

MEXICO CITY

In February 1932, the US Company's office in Mexico City received a wire from an unknown source and in a strange language, directed to the general manager of the local subsidiary, M. H. Kuryla. The verbs were in English, and some other terms, like "transit" and "Salvador," seemed familiar for anyone, but unlike any known language, most words started with *Z* and many others with *Y*. Kuryla immediately identified the language and used a sheet with translations. The telegram contained news regarding the evaluation of mining properties in Honduras (*Dali*) by the exploration department (*Zainetto*) and required assay supplies and equipment (*Zymotic*) for further development. The sender also announced his arrival via Mexico City (*Yecman*) in Pachuca (*Yggdrasil*) to discuss the company's expansion into Central America. *Zebrawood* signed the message, the coded name from C. F. Moore, who was at the time the vice president of foreign operations of the US Company in Boston, coded as *Zahurda*.[35]

The coded wires between top officials of the multinational corporations were just one manifestation of layers of internal information that circulated within this underground titan. The operation of a large multinational firm like the US Company required the sharing of information, capital, and technicians inside the corporation as a way of protecting internal decisions, estimations, technology developments, and the company's very future. They did not code everything with *Z*s and *Y*s. Geologists, metallurgists, and mechanical engineers designed, interpreted, and guarded blueprints of mines, machines, and plants; accountants, auditors, and managers generated, calculated, and deciphered coded sheets with cost estimations, reserves, valuations, and profit projections; labor experts, physicians, and lawyers shaped, explained, and booked reports, medical files and legal concessions describing labor policies, health records, and property rights. With no watchmen outside the firm, the custodians of the corporation established an auto organization. They designed the system of their power: the encrypted blueprint of their relations.

The construction of this techno-political body implied the recruitment and transfer of staffers, but also the production of new officers and, finally, the exclusion from the ranks of external elements. As analyzed in chapter 3, the US Company financed schools and universities, which, in turn, formed technicians specialized in corporate needs and embedded in the coded language of big technological mining firms. This was not exclusive of the Cambridge elite schools. The United States Lead Refinery, in East Chicago, had strong relations with local universities in Illinois and Ohio, organizing practices and internships for chemical engineers and operations design. The company's Utah division gave grants for research on western mining at the University of Utah. In Alaska, the Fairbanks Exploration Company was vital in creating the Alaska Agricultural College and School of Mines. Company officials were often teachers at universities, and the company hired undergraduate engineers with flexible plans. The Mexican division also had relationships with local technical schools in the 1930s. The company entered an agreement with the Escuela Prevocacional de Pachuca of the Public Education Secretary (SEP), establishing practical internships in the workshops of the Maestranza for the students. In addition, the firm offered visits to the mines and plants for professors and students from the engineering school of the National University.[36]

Despite these programs, the actual employment of the company was still very segregated. Even if the corporation had strong relationships with the local university, the Alaska-educated engineers did not participate in the global career paths available for the engineers of the East and West Coasts, even if the inverse path was very common. Moreover, in the Mexican division case, the company refused the employment of Mexican engineers from top ranks, despite the long-standing tradition of Mexican mining engineers. Since 1925, the mine department of the Mexican government required mining companies to hire engineers from the Escuela de Minería to supervise their safety regulations. By that time, only around 5 percent of the mining engineers from the National University found employment in mining firms.[37]

The exclusion of Mexicans from top ranks within the US firm had one exception: the legal team. Like most foreign interests in Mexico, the US Company established lengthy relationships with Mexican lawyers, who acted as representatives of the company before Mexican authorities. As legal processes are nation-specific, highly dependent on local knowledge of law, culture, and customs, only Mexicans had the necessary expertise and local political relationships to fulfill their obligations. The legal representatives of the firm in different periods—Salvador Cancino, Pablo Martínez del Río, Carlos Sánchez de Mejorada, and Luis J. Creel—were all members of prominent families in Mexico. Moreover, although some of them did travel to Salt Lake and Boston on different occasions, their engagement with the company was not as strong as their American counterparts. When the general manager of the Mexican division, M. H. Kuryla, tried to gain exclusivity to the services of Sánchez de Mejorada in 1934, the other companies that he represented offered him the position as general attorney for the National Chamber of Mining (Cámara Nacional de Minería en México).[38]

Several factors can explain this nonconvergent technological gap. Some scholars speculate that one factor was the lack of advanced training in traditional mining schools, which prevented the local market from offering qualified experts to new high-tech companies.[39] That was hardly the case of the Mexican division, as Mexico had been the leading producer of silver and the Escuela de Minería had trained highly prized technicians for centuries. This tradition ended with the arrival of multinational firms and the cyanidation process in the twentieth century, when Mexicans could no longer compete with their United States counterparts, even after the state promotion of their employment.[40] The definition of authority inside the firm required sharing knowledge, but also a distinct identity. Cultural differences, racial prejudices, and the construction of networks in the Anglo world perpetuated a corporate identity linked to the origins of the firm. The consciousness of itself built the Leviathan as a global organism, but it thought in English only.

SALT LAKE CITY, UTAH

"Real cooperation among individuals rests upon simplicity, charity, forgiveness, and kindness," began an editorial of the corporate magazine of the US Company, *Ax-I-Dent-Ax,* in December 1930. The article, by Edgar M. Ledyard, the chief of the agricultural department located in Midvale, further explained the meaning of each of those principles, focusing on the cooperation inside the company. It was, in many ways, the articulation of the corporate spirit described by Jennings during those years. "Cooperation among members of a business organization is somewhat different than that which exists between detached individuals. Cooperation in an organization is reasonably expected; between individuals is entirely voluntary. The progress of an organization depends upon the cooperation of its employees and, upon its advancement, rest their future."[41]

This very technical "Christmas spirit" was not rare in the editorials of the magazines. *Ax-I-Dent-Ax* was an initiative of the safety department chief, Roy S. Bonsib, in the lead refinery in East Chicago, to advance the Safety First agenda of the company in the 1910s. Similar magazines appeared in other big multinational corporations at that time. In 1915, Anaconda Copper started published *The Anode,* edited by its safety engineer A. S. Richardson, and the same year ASARCO started publishing a corporate brochure.[42] After his success in the US Company, Bonsib applied the same strategies when he was as chief inspector of security in the New Jersey subsidiaries of the Standard Oil Company in the early 1920s.[43] Furthermore, *Ax-I-Dent-Ax* was not the only magazine produced by the US Company. Firms used different booklets to dictate company policies and regulate behavior within the multidivisional corporation. During the 1920s, the subsidiary US Fuel Company gave monthly pamphlets to coal dealers, *The Firing Line,* providing information about the products, prices, marketing strategies, and market opportunities.[44]

Born with a practical objective, *Ax-I-Dent-Ax* rapidly transformed into a corporate magazine for the company's top officials. It began publishing best practices in the workplace, strategies for foremen and managers in implementing them, and the results of the different subsidiaries of the corporation. At the bottom of the pages, it included security tips, accident descriptions as examples of bad behavior, and, little by little, motivational phrases for the employees. Soon, the magazine became the principal vessel of the US Company's identity. The cultural shift became clearer after Bonsib left the company and the magazine started to be printed in Salt Lake City, by the chief of the agricultural department. Every issue had an editorial subject focused on the operations of the company, the seasonal holidays, or the political scene. For Edgar M. Ledyard, "*Every Ax-I-Dent-Ax* is really a civics number, for each one contains something about the social relationships of employees or safety to mechanical developments or the future possibilities of the Company."[45] The magazine, published centrally, promoted a universal morality among the staff and highlighted the importance of loyalty to the corporation. It also revealed the conception of collective unity within the firm, the Underground Leviathan. "While a company is an impersonal thing," wrote Ledyard, "it should have the same consideration from an employee that the employee expects from the Company."[46]

The magazine hired correspondents in every division, who reported every month on the lives of the company officials. Nonetheless, not everything was indoctrination, as most of the pieces were written by company officials on a variety of subjects from their shared experience. Many company officials submitted reports demonstrating the importance of the magazine in the lives of the staff of the multinational organization. In its pages, engineers portrayed the technical difficulties of new processes, printed blueprints of designs patented by the company, discussed the possibilities of international metallic markets, and considered the social, mining, agricultural, and industrial conditions of the localities around the company's many divisions and subsidiaries. Thus, the magazine reflected the absence of a clear line between the corporate objectives and those of its engineers. The multinational was not only their employer but also they themselves were a body of self-regulated technicians.

The technicians of the firm were not alien to the sharing of international technical experiences. The expansion of the activities of American mining engineers was imbricated with the growth of multinational mining companies to the Global South, being themselves the designers of the new models of industrial mining. Mobility experience and identity with corporations, within a profoundly white male middle-class profession, created a keen collective consciousness of a transformative endeavor. Mapping, knowing, and redesigning the New World were shared traits to the experience of this group.[47]

Nonetheless, while magazines such as *Engineering and Mining Journal* or *Mining World* reproduced this ideology of design, the corporate magazines reflected another type of identity, more restrictive to a single company yet more open to other top staffers. The corporate brochure gave news on the integral outsider life of company officials, from visits and trips, vacations, diseases, births, deeds; marriages of engineers, managers, doctors, nurses, assayers, and clerks of the corporation; their new cars, picnics, and the experience of new gadgets in their shared consumption promoted by the United States Stores. Correspondents published photos documenting the operations of factories and office gatherings, as well as jokes about the lack of efficiency in some of them. Employees wrote in the magazine short stories about their previous employment, original papers on political subjects, and accounts of long trips or little vacations. In the magazine, an employee could read the history of mining in Scandinavia by the research department official, G. N. Kiserbom, or about juvenile employment on a cargo ship in Texas from H. B. Allen; the road trip of the employees to Acapulco or the vacation of Safety Inspector James A. Cotter to Yellowstone; the experiences of O. Henry and the US Exploration Company team into El Salvador in search of mining opportunities, or the witty comments of western engineers on their trips to the New York divisions.[48]

This community construction also shared the necessary portrayals of otherness: the society outside the corporation and the natural environment often mixed one with the other in the accounts in *Ax-I-Dent-Ax*. Descriptions of the relations with natives in Alaska, especially during fairs and contests, always exalted their physical capacities into the unbeatable nature. In the South, only American officials of the Mexican division published observations and opinions of Mexican traditions, such as the Catholic Christmas, the posadas, the Judas burned, and the Day of the Dead. An official like A. B. Marquand, general manager, would write on all sorts of subjects, from the difference of the monetary systems in Mexico and the United States, the life of the Anglo community in the capital, and the exploration possibilities of southern Mexico and Central America.[49]

The conquest of an unregulated space was another part of the collective experience of these experts. This space not only appeared in the tropics but also in the divisions in the West and North.[50] Employees shared their strategies of snake elimination inside the Utah mines, and problems with bears were mentioned by the watchmen of the Davidson Ditch in Fairbanks, along with the different species of mosquitoes found in the hot summer in Nome. In spring, *Ax-I-Dent-Ax* published comparisons of the agricultural achievements of company officials in other latitudes. Their homologated life and consumption was always related to the technological endeavor of top-ranked officials. The director of the agricultural department gave scientific tips on house gardening, and the researcher of the lead refinery provided advice on the different chemical compounds used in the cleaning of cars. In a word, planning mediated experience, and every problem or contradiction was a designing puzzle.[51]

Science, technique, conquest, and control were central to these narratives, from the stars to the bottom of the earth. Officials published graphs showing annual temperatures in the different subsidiaries, seasons of growing and crops in the territories, and inventories of indigenous vegetables and flowers, or made descriptions of the volcanoes in Mexico. Technicians explained the experience of the months without light or darkness in Alaska, the optical effects of the snow refraction known as parhelia, and the midnight sun of the Arctic pole.

Indeed, the publication revealed the scientific relationships of the engineers and the firm with its local contexts. The subsidiary Fairbanks Exploration Company (F. E. Co.) collaborated in the construction of Polar Year Magnetic station, with the Alaska Agricultural College and School of Mines, two stations collecting magnetic data and one measuring atmospheric electricity, publishing reports in the corporate magazine. In Mexico, the general manager M. H. Kuryla was one of the promoters of the measuring of cosmic rays in collaboration with the University of Chicago.[52] *Ax-I-Dent-Ax* even printed a description of three-foot fossil prints of what was considered a dicynodont, found in the coal Panther mine of the US Fuel Company in 1929. In Alaska, the washing and dredging of permafrost gravels uncovered so many fossils that the company started collaborating with the American Museum of Natural History in the early thirties. The more important finding during the company operations was the mummified head and leg of a young woolly mammoth, with skin and flesh, and a human-made artifact. The baby mammoth was named Effie, in honor of the F. E. Co.[53]

The appropriation of nature reorganized by the technical staff of the firm expanded in the cultural, symbolic, and scientific views on the company operation. The growth of this corporate community into different spaces implied a process of corporate discovery. Company officials' interests were not restricted to the ores, nor solely to the economic results of the corporation, but also in the deeper meanings of their endeavors. Some staffers constructed a consciousness of the historical character of the firm, linked to the earth they transformed. These mining engineers were some of the first historians of the different places they occupied.

Edgar Ledyard, the chief of the agricultural department in Midvale, Utah, was the president of the Utah Historical Landmarks Association, as well as a member of the Utah Academy of Sciences, of the American Geographical Society, and of the American Association for the Advancement of Science. As editor of *Ax-I-Dent-Ax* after the departure of Bonsib and until his own death in 1932, Ledyard published papers on the history of the Utah territory and the role of mining, especially on the foundation of Salt Lake City. Along with R. T. Walker, geology chief of the corporation, the historian Charles Kelly, and the archaeologist Julian H. Steward, he traveled and photographed the Mormon and gentile trails from the East to the West. The study was a first in regional archaeology, documenting old Mormon settlements and rest stations, and interviewing the descendants of riders on the Pony Express Trail, Oregon Trail, and Santa Fe Trail. In the history of the territory by these mining experts, there was no place for the indigenous or Mexican populations that inhabited the space for centuries. The only history was the account of white control over the West, and they drew a continuity between the Spanish settlers, the Mormon farmers, and the corporate power in the valley. In subsequent years, Ledyard continued recording routes to Utah used during the 1840s, writing a history of the settlements in the territory since the Spanish and the early conflictive history of occupation of the Great Basin. He was particularly interested in San Juan County, collecting photographs, diagrams from travelers, and oral histories. Finally, Ledyard identified the history of the territory as ultimately leading to the US Company itself. In researching over the origin of industrial mining in the area, he discovered that the first claim in the copper mountains of Utah was now part of the Lark mine of the US Company.[54]

All the roads led to the corporation for these engineers. George W. Cushing, head of the traffic department, wrote in 1929 a general history of communications in Utah, from the Pony Express Trail to the construction of the traffic department of the US Company.[55] In Alaska, John Boswell, who came to be the general manager of the F. E. Company, worked very closely with the Department of Anthropology and Archaeology of the recently founded University of Alaska Fairbanks (UAF), helping in the recovery of artifacts found during the stripping of the ground. In 1935, he was president of the UAF Alumni Association and created the Boswell prize, awarded to female seniors. Twenty years later, Boswell was a delegate to the Alaska Constitutional Convention, part of the executive committee, and started to write a history of Fairbanks. Boswell's narrative portrayed the company history interwoven with local urban and agricultural history, as well as the foundation of the university. His book, published after his death in 1978, was dedicated to the officials that first employed him in the company: "I suppose we could call this my last report."[56]

As with other aspects of the operation of the firm, the technical crew in Mexico had a robust identity relationship with the corporate transformation of the surroundings. Alan Probert was the best representative of the scientific and historical interest of officials in Mexico. A Berkeley alumnus, he had worked in several divisions of the US Company in the West during the late 1920s and 1930s, until he took the position of manager in Mexico in the early 1940s. One of the officials more interested in learning Spanish, he read the texts of Alexander von Humboldt and studied the archive of the company dating to the eighteenth century.

Similar to Ledyard in Utah and Boswell in Alaska, Probert delineated a continuity between white expansion, technical expertise, corporate power, and himself in the centuries of exploitation of silver in the Pachuca and Real del Monte region. During his years in Mexico, Probert visited the Archivo General de la Nación in Mexico. After the Second World War, he transferred to Alaska, where he recorded traditional and modern methods of mining. After his retirement, and still obsessed with the Mexican division, Probert visited the Archivo General de Indias in Sevilla, the Huntington Library in California, the

Library of Congress, and traveled to Cortegana, the birthplace of the Count of Regla, the founder of the Compañía de Real del Monte, interviewing the descendants of the noble. With the materials collected, he wrote biographies on the inventor of the amalgamation process of silver extraction, Bartolomé de Medina, of the Counts of Regla, and of the longtime manager during the British company, Antonio Buchan. Probert had lived in the old house of the English manager, as it was a property of the US Company, and had the collection of painting of Mexican landscapes by the Italian Eugenio Landesio.[57] Probert saw how his character as an Anglo engineer, an integral outsider, and a company official transcended time, reaching back to Buchan. "The landscapes of Real del Monte were memories for us, as they were for their first owner, of the fabulous mining district of Pachuca."[58]

LENINGRAD

In February 1920, a year after he listened to F. W. Draper's presentation to the American Institute of Mining and Metallurgical Engineers in Boston, Sidney J. Jennings, vice president of the exploration department of the US Company, announced the transformation of the organization into the American Institute of Mining and Metallurgical Engineers. The new name of the association reflected the expansion of the professionals operating in the mining business. During the year he presided over the professional association, new members were integrated from China, Japan, Korea, Sweden, Australia, Chile, Mexico, Canada, Italy, and Nicaragua. Jennings had negotiated a permanent agreement with the Canadian Mining Institute and calculated, in his speech, around fourteen thousand members, associates, and junior associates of the organization who extracted minerals in every corner of the world. He described his time as president as a revolution of mining and metallurgical engineers in the world. "We are living in strenuous times. Governments are being overthrown every day," said Jennings, "so it is not strange that the Institute itself has decided upon a radical, nay a revolutionary step."[59]

The new president, Horace V. Winchell, was much more concerned about the role of the engineers in the revolutionary countries: Mexico and Russia. "How many of our engineers are now enjoying princely honorariums in Russia?" he asked, "None. How many are at work in Mexico?" he continued, "Very few."[60] Like many in those years, Winchell considered peasant and worker leaders the most dangerous antagonists of the educated engineers and urged his colleagues to take a more prominent role in politics. The effective resistance against the socialist revolution was for technicians to take power directly.

The members of the AIME were already on it, as the war had involved them in planning for other sovereigns. In the following years, mining and metallurgical engineers increased their influence in large organizations, not only in the corporations but also in state-controlled firms. In Bolivia, American engineers were vital in the tin boom of 1920 and the reconstruction of the mining enterprises after the crash of 1929.[61] After the rebellion of Haitian cacao farmers against Oreste Zamor, the new president of Haiti, Davilmar Théodore, offered prospecting rights to US companies and to pay the salaries of American mining engineers, in exchange for government recognition.[62] Another mining engineer, and member of the AIME, started a promising political career after World War I in the United States. After spending most of the decade as the US secretary of commerce, Herbert Hoover started his presidency in March 1929. Hoover portrayed himself not as a politician, but as a problem solver and a technology enthusiast.

During the first decades of the twentieth century, engineers working in mining corporations had tremendously increased their capacities of planning over exploration, production, distribution, and consumption of minerals, and their political capabilities became evident on a national and international scene. They had transformed the underground world, materializing and fueling, in very literal terms, the dream of an industrial planned economy. They were not driven only by manifest destiny, nor only by the profit search of their shareholders. Staffers and experts created an organization, the corporation, which in turn shaped them and shaped the world. As subjects, they were the architects of their collective iron cage but, moreover, of the blueprints of other actors. They had created their own confederacy of technicians.

The trajectory of Leslie Douglas Anderson can better exemplify this relationship between technical designs and the engineering of political relationships. He was a mechanical engineer from Michigan and a member of the American Society of Mechanical Engineers, the ASME, and the Utah Society of Engineers. Anderson started working for the Midvale plant of the US Company in 1906 and followed the corporate career ladder to the general headquarters in Boston. He had defended the company in pollution trials in Salt Lake in 1916 and had surveyed some more important technological advances of subsidiaries in Mexico, Utah, and Alaska. He designed "the largest flag in the largest pole in Utah" for its installation during the First World War. In one of the more multiethnic and patriotic gatherings organized by the company, Anderson agitated the crowd with a speech over the defeat of the enemies.[63]

In December 1930, *Ax-I-Dent-Ax* reported that Anderson had left his position as chief of engineers in the US Company. With the metal prices at a historic low, Anderson worked as a consulting engineer for the Freyn Engineering Company of Chicago, specializing in the design of infrastructure. This temporary position allowed Anderson to maintain his employment inside the US Company, with the subsidiary United States Lead Refinery. During his trip to Australia, Anderson did not miss the chance to write to the corporate magazine about his observations of the railroad system and its integration into the industrial development of the country. When he took a ship on November 25, 1930, nothing appeared different. His experience as a mechanical engineer had taken him from Michigan to Arizona, from Chicago to Salt Lake City, and from Alaska to Mexico. "A biography such as Mr. Anderson's should be inspirational to the student as well as to the graduate engineer," printed *The Michigan Technic* after his departure in 1931.[64] This time, Anderson would retrace the trip of Draper fifteen years earlier, but in the opposite direction, moving to Kuznetsk, Siberia.

The return of US engineers into the socialist East was rather fast. In 1923, some firms in the platinum business employed, again, mining engineers in the Ural Mountains.[65] Although under the New Economic Policy of Lenin private capitalists could reinvest in the socialist economy, it was not until the first Five-Year Plan of Stalin that American engineers traveled in considerable numbers to Russia. The industrial effort of the Soviet Union in those years heavily depended on the importation of goods, supplies, and technology from the United States. At the peak of this transnational trade, US imports represented up to 25 percent of the total imports of the Soviet Union.[66] This material flow reflected an ongoing technical collaboration with US engineers. At least two dozen US engineering firms reached agreements for technical assistance in an array of industrial sectors, from meat-packing to automobiles, irrigation to power generation, and the production of fertilizers to pencils. In exchange, engineering students were allowed to attend Harvard and MIT, even before the official recognition of the Soviet government by the United States.[67] In this vast and extensive partnership between technical elites in both countries, mining and metallurgical engineers were especially vital to the first major Soviet plan. In particular, American engineers were critical in the construction of the coal, iron, and steel facilities in Siberia.[68]

In 1927, the Freyn Engineering Company of Chicago won a concession for technical assistance to the State Institute for Projecting Metal Works, GIPROMEZ, in Leningrad. The firm would evaluate the existing plants and would design new ones according to the needs of the already approved Five-Year Plan. On August 2, 1928, Freyn received a second contract, covering the engineering, design, and superintendence of construction of an iron and steel plant with a 1.5-million-ton annual capacity.[69]

The project in Leningrad employed around thirty engineers, while seventy more worked for the construction and supervision of the iron plant. None of these engineers were strangers to corporate planning and social engineering; none of them found anything new in working for a socialist economy. For Anderson, the experience was not much different from the big projects in Alaska or Mexico. The firm and the state controlled their housing, offices, food, supplies, leisure, and travel. The meat, vegetables, and other food products were provided from the cooperative stores, and several Russian domestic workers prepared their food and arranged their houses. The planning of the corporations, on one side, coincided with the hierarchical regulation of the Soviet Union. As in Mexico, Peru, Alaska, or El Salvador, the technicians in the Freyn firm interacted in local contexts with translators, lower-rank technicians, and domestic workers. However, on the other side, it clashed with the socialist ideals of some inhabitants of the old winter capital of the empire. In a heavily publicized incident, one cook of the Freyn Club in Leningrad faced down a crowd of women in the line for the cooperative stores. As usual, the employees of the *Amerikanski* could skip the line to get the goods for the new US community, but the conflict with the protesting customers had to be dissolved by the manager of the socialist store and three militiamen.[70]

The steel mill, named the Stalin plant, depended on the coal ores of the Kuznetsk Valley provided by the state mining enterprise *Kuzbassugol* (Coal Kuzbass), which functioned as part of the gulag system of Siblag. A veteran in the design of lead and copper furnaces in Salt Lake, Anderson moved around the Soviet Union, observing successful refining methods of Russian ores. He still reported regularly to *Ax-I-Dent-Ax,* and the company community celebrated his engineering achievements. Anderson made a brief visit to the lead plant in East Chicago in the summer in 1932, and two of his sons visited him the next fall in Siberia. He was not the only former employee working on the project. Bennet Stewart also reported about the traditions, music, and accommodations for US engineers in the Soviet plant's construction, accompanied by photos of the surroundings.[71]

By the end of 1932, Anderson was one of the five engineers that remained on the project and continuously sent postcards to old colleagues of the US Company. In what the editor of *Ax-I-Dent-Ax* thought was the most remarkable account, Anderson described a photo of the fountain of Winter Palace, taken by the Bolsheviks fifteen years before. It was no longer the St. Petersburg of the first American colony, where Draper had accumulated platinum for the war effort. It was by then a different city, named after the late leader of the Bolsheviks. Anderson could feel, as well, the power of the sovereign body in the fountains of the palace: "Peter the Great's dream of Versailles in Russia, 18 miles from Leningrad on the bay. Warm regards."[72]

The success of the consulting firms for the first Five-Year Plan (1928–32) informed the strategic decisions of the second Five-Year Plan (1933–37) and the construction of even larger metallic-treating facilities farther West. New Soviet mills and plants were built during the following years, with a combined projected cost of almost a billion rubles, and a forecasted production, by the last year of the plan, of 2.6 million tons of pig iron. In 1934, one of the star industrial complexes of that period, which could treat up to 500 tons of Kerch vanadium a year, was inaugurated. The facility had "blast and open-hearth furnaces, rolling mills, coke, and chemical works, and large sintering and flotation plants" to produce high-grade steel: the Azov plant at Mariupol.[73]

Anderson's trip abated the distance between Peter the Great and Stalin and from the company towns and plants in Pachuca and Utah and the gulags and plants in Siberia. It unveiled that the relation between the corporations in North America and the planned economies in the East was not of contrast, but a continuity. It revealed the deep consciousness of the transformative nature of engineering while developing an awareness of the engineer's place in corporate, planned organizations. By that time, staffers inside a corporation not only decoded documents, blueprints, projections, reports, strategies, and authority, but also functioned in a coded script of behavior. The code was, in one way, imposed by a systematic organization, yet appeared as an auto-created script, a blueprint of themselves and their relations, a Leviathan and as a body of engineers. The strong identity of corporate engineers and their planned experience of space and rule was not the only one inside the firm. By the 1920s and 1930s a very distinct voice disputed the conception of rule of the multinational corporation and the construction of space in the mining district. This actor reached a local and international space: it organized an international body of workers.

CHAPTER FIVE

An Autopsy in Every Case

A prince ought to have no other aim or thought, nor select anything
else for his study, than war and its rules and discipline; for this is the
sole art that belongs to him who rules.

—Machiavelli, *The Prince*, 1532

JOPLIN, MISSOURI

"There are mean things happenin' in this land," said Sheldon Dick, the
narrator, and the producer (along with Lee Dick), of the short documentary *Men and Dust*. "There are mean things happenin' in this land,"
repeated the refrain of the film, presented in 1940 during a series of
interventions in the Tri-State Conference celebrated in Joplin, Missouri.
"There are mean things happenin'; there are mean things happenin'"
continued the refrain, showing footage of the conditions of living in
the mining town, one of many in the lead and zinc area between Oklahoma, Missouri, and Kansas. The Dick family was in the city as part
of the National Committee for People's Rights, which organized the
event to share some of the results of two years' work in the area with
the attendance of the Roosevelt's secretary of labor, Frances Perkins,
and the retiree Harvard professor Alice Hamilton.

"Lead for pipes, for bullets and gases; zinc for medicines and batter-
ies and paint. Thousands of tons of rock poured daily, 20-million-dollar
business every year," explained the documentary. The area produced
10 percent of the lead and 38 percent of the country's zinc, and the
industry had grown in recent years due to the increased demand for
the war. It was not a new mining area. During the 1910s, mining in lead
and zinc exploded, especially during zinc's price boom during World
War I. In 1915–16 the US Company alone installed three refineries in
Kansas, in Iola, Altona, and La Harpe; one refinery in Oklahoma, Checo-
tah; and three mines in the area: Ravenswood in Missouri, Naylor Tract
in Kansas, and Ritz in Oklahoma. The mill and the refineries installed

by the Ravenswood mine, powered with natural gas extracted on-site, could refine over six hundred tons of concentrates every day. The refineries' capacity and cheaper cost allowed the firm to process zinc from other company properties, especially in Mammoth, California, and Leadville, Colorado.

Nonetheless, soon after the end of the Great War, the district experienced a long-standing depression. Most of the firms either scaled down their operations in the tri-state area or rented their properties for local entrepreneurs, waiting for a recovery in the prices. A second boom in the area arrived by the end of the 1930s with the extensive war demand for zinc, and the corporations were ready to take back the properties. By 1938, the US Company had restarted the operations under their rule, following the area's profitability.[1]

"Father has a B card, a C card, a D card; Father got an F card, Father has the Roop; a silicosis F card that won't let him work," continued the narrator in *Men and Dust*. The committee had succeeded in bringing Perkins to the area and arranged for her to participate in a hearing on the Senate during the summer. Perkins was the first-ever woman on a presidential cabinet, the most stable member of Roosevelt's team, and one of the key actors in the government's dialog with unions. In 1936, she started a robust campaign with the National Silicosis Conference organization, inviting union and industry representatives. The next year, Perkins toured mines, quarries, and granite factories, and released an important information campaign. The film *Stop Silicosis* detailed the risks of the disease, its pathological evolution, and the different marks it left in the worker's body: the small scar tissue left in the alveoli, as was revealed in the X-ray prints. The short film showed the fate of John Steele, a worker in a quarry who, after developing early signs of silicosis, is laid off from work. Weak, he finds a job in a tombstone workshop, where he engraves his name.[2]

"All he needs is to quit the mines forever, to escape forever the knife-sharp flint dust. All he needs is years of rest, years of sunshine, and milk and rest. Well, he's quit the mines forever. He will die when the sap flows. He will be dead in the spring"—continued the short film. In those years, the labor department considered that federal regulations were unnecessary, especially in the mining operations. Some states approved compensation laws, and corporations had widely adopted wetting during drilling in the underground and increased ventilation in the tunnels. Nonetheless, by 1940, most of the medical examinations on workers were only available as a form of corporate control over workers' bodies and not as a mechanism of access for the workers to see the mark dust left in their bodies. The imprint of the work in their lungs was available for the firm but not the worker. Moreover, the Public Health Service agreed, in most cases, with the employer's doctor. In an exchange between Perkins and the local public health official, Dr. Sayers, the latter assured the listeners that silicosis was not a source of disability and that it should not be compensated as such unless it appears in its most extreme form. Moreover, he argued that it was complicated to assess if the damage was a product of the work or infection, that is to say, tuberculosis. During the event in Joplin, Perkins estimated local officials' demand to identify damage in the lungs as forensic geology. "I am sorry to say that what you really want is an autopsy in every case," she replied after Sayers's medical exigencies.[3]

"There is nothing dramatic about sickness on the Tri-State. Only, somehow, always somebody has something. The mother has a fever; Gene has tonsilitis, Mary keeps her cold," Sheldon Dick narrated. Alice Hamilton, another attendee at the conference in Joplin, returned to the area after having conducted the study over the country's lead industries twenty-five years earlier. In extensive surveys during the war, Hamilton recorded some of the diseases provoked by dust in the mining methods. However, her primary focus was on the effects of metals and not silica in the miners. Like most of the physicians at the time, she believed that the installation of wetting and mucking in drilling and shoveling operations was enough to prevent the disease and attributed some of the failures to some of the workers' stubborn culture. Nonetheless, the disease's persistence signaled to the seventy-one-year-old doctor the flaws in controlling and preventing the prenetration of dust into the lungs. She saw a familiar landscape to the one she witnessed twenty-five years before, except that the tailing piles were bigger by the day. The mine was extending outside the gate of the firms, with a dust that invaded the town. Moreover, the damage in the lungs of the miner expanded in the community as an infection. Women in the mining towns were almost as susceptible to die of tuberculosis as the working men, and around 35 percent of the children in the communities in the tri-state area tested positive for the bacteria. The dust had to be controlled, she claimed, in the mining operation, the surface plants, and housing developments.[4]

"The people live in the shadow of the tailing piles. Tailing piles are crushed rock, rock from which the ore has been extracted. Rock to gravel, to sand, to dust. And the dust blows, over the towns and the streets, the street and the houses, the houses and the people, the school and the children. They live in this shadow," continued the voice in the *Men and Dust* documentary. Despite the film's name, silicosis's problem revealed a forgotten entity in the miners' culture: the women. Most of the underground work and in the plants was carried out by men, but women started to appear as clerks and secretaries of the mining operations. Moreover, migration patterns had changed in the mining business, and the share of lone men that came back and forth between the mines and farming diminished. More families migrated to the camps, and the women, as in Hamilton's analysis, were suffering from the same diseases as the working men in the mine. The women were not only the object of disease in the mining camps but also were a central piece in the organization of labor in them. Perkins and Hamilton were two of the leading representatives of the growing role in labor regulation by women. However, women became increasingly central in some of the unions disputing the expansion of management control into their communities.

"In the meantime, they sicken; in the meantime, they die. Charlie, and Roy, and Norman, and Mac, and Wilbur, and Alice, Chester, and Tom, Betty, and Anne, and Dewey. Somehow, somebody always gets something." The short film showed the pictures of some of the community's late members that died after the committee's research. Another attendee portrayed the same faces at the conference: Norah Hamilton. Norah engaged in the Hull House for decades and joined Alice in some of its political trajectories. Alice traveled in 1924 to Soviet Russia, as part of the League of Nations' health committee, and had documented the work against typhus in 1921 by the Soviet government. Three years later, Norah made the same journey. More than her sister, she was critical of the state's control over the workers' decisions and the control over the political activities inside the party, and she denounced the dissidents' disappearances. In the next years, they were involved in several civil liberties organizations. Norah and Alice Hamilton participated in protesting against Sacco and Vanzetti's case, defended conscientious objectors to war, joined the anti-fascist organization Lincoln's Birthday Committee for Democracy and Intellectual Freedom, and participated in the American Committee for the Protection of Foreign Born. This time in Joplin, in 1940, they were both members of the National Committee for People's Rights, which had organized the conference, and the Tri-State Committee. Alice's younger sister sketched the faces of the town's tragedy in the book *People of the Wasteland.* As her sister had in the surveys, Norah visited the miners in the hospital, their houses, the workplace, and the union meetings. The grainy drawings depicted the invisible dust through the community members' expressions, and she seemed particularly interested in the figure of the leader of the International Mine and Mill Union, Reid Robinson.[5]

"Give us health, give us work, give us life. Give us health! Give us work! Give us life!" concluded the film *Men and Dust,* this time with the voice of the union leader, Robinson. Like the Hamiltons and Dick, he was part of the National Committee for People's Rights and attended the conference in Joplin. His relationship with the Hamiltons was strong, and Alice suggested the program of the union, which demanded federal workman's compensation and federal minimum labor standard laws that complemented the state-level laws, a cheap housing development away from the tailing piles, and a regional sanitarium to provide health care for tuberculosis patients in the tri-state area. The demands included that the state as a sovereign power should take care of the worker's body: a legitimate authority in regulating the different stages of the working life that emerged from the mine. The organizers also were the product of decades of conformation and redefinition of workers' bodies since the strike wave of 1919–23.

This chapter argues that the definition of an organized body of workers and the appearance of the workers' bodies in the public debate went from the local to the national and then to the international arenas. While the years after the First World War were marked by a diversity of successive local actions of unions, the years prior to the Second World War witnessed the emergence of an articulated international miners' movement. This new body of workers consciously followed the different branches of the firm's sovereignty and created an autonomous yet convergent organization.

On September 21, 1940, the federal relations board resolved a dispute between the local 444 of the Fairbanks Mine Workers' Union and the United States Smelting, Refining, and Mining Company. The firm had hired an Alaska lawyer, Southall Rozelle Pfund, and Ropes & Gray LLP, a firm that recruited the Harvard Law alumni elite. This time, the firm had assigned Charles E. Wyzanski Jr., Archibald Cox, and Henry C. Moses to the case. Wyzanski had been, by then, solicitor of labor for the US Department of Justice, while Moses had been the assistant of the Republican senator Robert Taft. The representation of the union was much more modest, composed by the local union organizer, W. M. Rasmussen, and his local Alaska lawyer, J. A. Lathanan. The company disputed the vote of recognition of the union in its operations at Fairbanks, which represented the first collective agreement signed in the northern metallic operation in the continent. Simultaneously, it was the last of the firm's divisions to establish formal dialogs with workers' organizations.[6]

The firm had, by then, a growing experience of dealing and negotiating with workers. Part of the firm's strategy before consisted in forming an international body of workers and managing conflict inside the firm by expanding the limit of its sovereignty. Nonetheless, since the strikes of the 1920s, the firm tried to push their favorite unions forward and maintain competition among the several organizations. In Pachuca, in 1926, the Confederación Minera Hidalguense, member of the Confederación Regional Obrera Mexicana (CROM), negotiated the collective contract #2 with the United States Company, the Santa Gertrudis Company, and the San Rafael Company over the workers in the district of Pachuca and Real del Monte. The new organization replaced the Unión de Mecánicos Mexicanos (UMM) by including workers from the underground, against the qualified mechanics of the Reglamento. The new agreement also revealed the distance that the UMM took from the leadership of the CROM. While the different railroad unions had radicalized and influential communist leaders took the locals' direction, Luis N. Morones, the general secretary of the CROM, established a long-standing alliance with the federal government and a politic of conciliation with owners.[7]

The idyll did not last long. In 1928 the CROM was accused of being responsible for the assassination of the elected president, Álvaro Obregón, and survived by widening its alliance with the companies' management. Later that year, the union only negotiated the extension of contracts with the firms in Pachuca and Real del Monte. Simultaneously, the UMM increased its strength among the rank-and-file workers on the surface and underground through another organization, the Alianza de Trabajadores Mineros de Pachuca y Real del Monte. The Alianza started to recruit company contractors and their employees, making most of the mine's maintenance work, breaking the traditional rivalry between the surface and underground workers in the industrial mining system. In September 1930, the governor of Hidalgo, Bartolomé Vargas Lugo, a former worker of the Santa Gertrudis Company, called for a convention to establish a national miners' union. In the coming months, the Alianza started to organize small strikes demanding wages increases and eight hours of collar-to-collar (counting from when the crew changed into working clothes to when it changed to the civilian ones) before the workers left the facilities. The approval of the Federal Labor Law (LFT) in 1931 conceded closed shops and provided special regulations for industrial unions. While the Confederación was still able to renovate the contract with the companies, the Alianza in 1932 presented a new project. The following months, the two unions fought on the Federal Labor Board (Junta Federal de Conciliación y Arbitraje) to keep the companies' negotiation status.[8]

The Confederación maintained contract #2 with the companies in the round of 1933, but the support among workers was virtually nonexistent. In contrast, the Alianza started the efforts for national articulation. With Governor Lugo's support, the leaders toured the country and signed other locals for a new alliance. On May 1, in Pachuca, over twenty-seven unions and twelve thousand workers founded the Sindicato Industrial de Trabajadores Mineros, Metalúrgicos y Similares de la República Mexicana (SITMRM). The locals were usually parts of smaller crafts unions linked to the railroads, and the founding of the new union was parallel to the erection of the other industrial union of the country, the Sindicato de Trabajadores Ferrocarrileros de la República Mexicana (STFRM). In short, the two new unions entailed the disappearance of crafts organizations into massive industrial structures. The locals 1 and 2, from Pachuca and Real del Monte, were confirmed by the locals 36 and 43 of the UMM, the Alianza de Trabajadores Mineros, and the local 38 of the Alianza de Ferrocarriles Mexicanos. The convention elected an old worker from the United States Company, Agustín Guzmán, as national general secretary. The next year, they organized thirty-six strikes with the participation of eighteen thousand miners.

Another relevant section for the US Company was the newly formed local 9, in Hidalgo del Parral. The firm had expanded in the area in the previous years, and the operation was the third largest in Mexico by the US Company. The Union de Mecánicos, Electricistas y Similares de Parral and the Sindicato de Trabajadores Mineros Benito Juarez had constituted section 9 in the foundation in 1934. That year, along with the local 11 in the neighboring community of Santa Barbara, the miners at Parral participated in a strike that demanded subsidized housing, transportation, and price control for essential products.[9] In 1934 the union had most of the mining contracts in the country. After some resistance, it integrated into the *cardenista* union system of the Confederación de Trabajadores de México (CTM) in 1936.[10]

The contestation of control of the firm's international body appeared everywhere in the firm's different divisions. Just as in the 1920s, labor unrest emerged from the underground, but this time the unskilled workers were the major force in unionization. Now, in the 1930s, the division in Utah tried to control labor unrest organizing, a common strategy with other corporations in the territory. In 1934 the ASARCO, Silver King Coalition Mines Company, Utah Copper Company, Utah Delaware Mining Company, Utah Consolidated Mines Company, and United States Company refused to implement the eight hours collar-to-collar regulation from the Industrial Commission of Utah, taking a collective defense to the Supreme Court of the state and, after losing the legal battle, simply refusing to comply. For its part, a new union took control of organized workers in the area: the IUMMSW.

The IUMMSW had a more radical agenda than its predecessors. It was one of the founding organizations of the Committee of Industrial Organizations (CIO), and, in 1936, elected Reid Robinson as general secretary. Red Robinson, as he was known, started an aggressive affiliation campaign in the West, and one of the main scenarios of their comeback was, again, Utah. Between October 9 and 15 of 1936, the locals in Utah of Bingham, Lark, US mine, Tooele, Tintic, Centennial Eureka, and Park City walked out demanding a union shop, wage increase, and the eight hours collar-to-collar day. The strike affected the biggest mining companies in the regions, Kennecott, Anaconda, and several divisions of the United States Company, which registered a total walkout of over two thousand workers. In the following days, the strikers won the support of the Utah Federation of Labor and the United Mine Workers Association (UMWA).[11]

The first mining industrial strike managed to paralyze the whole metallic mining operations in the state. The walkout's massive character and the strike's central organization reduced the conflicts and erased the confrontations with strikebreakers. Moreover, as with the conflicts in Pachuca, the government changed its attitude toward the strike. The director of the Industrial Commission, William M. Knerr, was involved in bargaining, and just one week after the beginning of the layoff, the governor, Henry H. Blood, offered his participation as mediator. By the end of the month, the negotiation process between the companies and the workers took place, for the first time, in the governor's office. It was not all rosy, but the bargaining process itself was considered a success for the union. Robinson arrived in Utah in December and accomplished a twenty-five-cent increase with Anaconda. Soon after, he started convincing the rest of the locals to accept similar agreements with the rest of the companies. By December 1936, all the canyon locals had signed collective contracts, their firsts, with Anaconda, Kennecott, and the United States Company. However, they still did not obtain a union shop.[12]

The unions' problems in organizing started to appear, at that moment, far north. In the summer of 1939, the International Mine and Mill Union decided that it was ready to attain unionization in the gold dredging camps of interior Alaska. The Fairbanks Exploration Company, the firm that dominated the farthest north operations in the continents, employed over 80 percent of the inhabitants of the town of Fairbanks and reported around one-third of the whole revenue of the United States Company. The firm was able, in little time, to identify the organizer among a load of migrant workers that arrived at the area every year: W. M. Rasmussen. He was quickly kidnapped by some company guards and put in prison. The limits of sovereignty between the firm and the local authorities were nonexistent in practice, and any reinforcement of the state's power only implied an increased role of corporate sovereignty over the territory. Rasmussen was able to leave jail in late November, when the season was over and the snow made Fairbanks unreachable.[13]

He started the organization of miners in the following spring, and by the beginning of the summer, most of the workers in the area were committed with the local 444 of the IUMMSW. On June 3, 1940, the Fairbanks Miners' Union filed with the nineteenth region of the National Labor Relations Board (NLRB) a petition to acknowledge their officers as representatives of the 550 production and maintenance employees. Nonetheless, by then, two other organizations disputed the representation against the division of the US Company: the Brotherhood of Alaska Miners, unaffiliated, and the Allied Craftsmen's Federal Union No. 22316, which was chartered by the AFL. The officers at the NLRB suggested the company and the unions make a consented vote instead of waiting for a direction of election by the board. In July, the unions organized the election. The unions obtained 54 percent of the vote, against 46 percent of the workers who did not want a union. From the different organizations competing, the local 444 of Fairbanks obtained 48 percent of the total votes cast. The NLRB officer advised having a second election between the unions, as the unionized option was the only majority of the first vote. The legal team of the firm, Wyzanski, Cox, and Moses, contested the election because they considered that a second election should contain the option of the IUMMSW #444 local or no union. As in Mexico, the dispute between the different workers' organizations kept the corporate rule over the camps.[14]

While the election was being disputed, the firm's group asked the NLRB to dismiss the case on several occasions. After many attempts, Alaska's local case was moved to the federal office in Washington, but it was already late August. The legal team of the US Company argued that the NLRB did not have the authority to order a runoff election. Moreover, even if they earned that authority, they were not allowed to omit the choice of "No union" in the second vote, since this was an attempt against the "freedom of choice" in the labor law and to the constitutional liberty of expression. After those arguments were dismissed, the Cambridge team of Wyzanski, Cox, and Moses asserted that, given the delay in the second vote, the second election should be held at the peak of operations next season, in July 1941. The oral arguments continued the next month, but the National Labor Relations Board resolved on September 21 that a second election with the remaining workers would be held.[15]

The defeat of the firm's Bostonian legal team pushed the election before the end of the year. In February 1941, the National Labor Relations Board certified local 444 as the workers' representative, but the union now had to struggle to gain a collective agreement with the firm. In May, the locals made the first strike vote in the area, and, after failed negotiations, the strike erupted on June 5. To the surprise of the union, local authorities and neighbors supported the strike, and the company was forced to sign an agreement three weeks later, for it was the peak of the season and the firm desperately needed labor. With the new contract, the IUMMSW controlled the biggest mining firms in the territory of Alaska, the US Company, and the Alaska Juneau Gold Company.[16] With the contract, the US Company lost the last of the not unionized divisions in North America. A new subject, an international worker, was constituted legally as an organization that could negotiate with the corporate Leviathan.

CIUDAD JUÁREZ, CHIHUAHUA

The massive dimensions of the Alaskan territory made communications between the company's workers very difficult, as the miners of a single firm could be as far as sixty miles away from each other. The companies usually controlled the town and the roads and telegraphs, and the workers migrated only for the working season. Only around 80 of the 750 workers needed by the Fairbanks Exploration Company had jobs all year; the rest were employed only between March and November. Year after year, locals' representatives had to maintain a campaign with the new arrivals, and they competed with the Independent Building Trades Union from the AFL. Moreover, delegates' life expenses were extremely high, as much as 220 percent higher than in Chicago, and they had to travel constantly. The delegate from Fairbanks had to organize the mines in Jonesville, over three hundred miles away; the gold mines of Kimshan Cove, an island eight hundred miles into the Pacific; and the dredges in Dawson City, four hundred miles into the Yukon in Canada. Most of the delegates were not prepared for the bulk of responsibilities in the territory, the hard work in the dredging season, the challenging weather, and the political prosecution. This created an exceptionally high activist turnover and constant financial disagreements between the national leadership and Alaska agents.[17]

The activities' concentration made the organizers an easy target for retributions, especially in the increasing militarization of the territory because of the war. Most of the organizers spent some time in jail, and, after the beginning of the war, they also faced repression owing to their nationality. The State Department had required mining firms to provide information on foreign elements in the mining camps. This new form of sovereignty, provided by the war effort, empowered the managers of the US Company, the main bearers of sovereign power in the area. The labor department at Boston sent, in December 1941, new regulations to the employment directors of the divisions. They should register all German, Italian, and Japanese employees, including birthplace, citizenship, relatives' names, and European connections. Additionally, they had to record "expressions or remarks from the individual employee under review of a disloyal nature," writing down the date, statement, place, and interlocutor. Finally, the labor departments of the firms had to register known communists: the "names of any employees known to be communists should be reported regardless of nationality or citizenship."[18] Soon, Ernest Schulz, one of Alaska's organizers who had lived in the area for twenty years, was apprehended. In the fall of 1942, the FBI accused him of being a Nazi, then a communist, and, in the end, of having supported the Spanish revolutionaries. "They called it acts against the country and government, enemy acts. Therefore I am to be isolated with a lot of other aliens for the rest of the war," he wrote to Rasmussen, the leader of local 444 in Fairbanks, when he was on his way to a concentration camp near Anchorage.[19]

The unionization of the workers in the mining camps farther up north in the continent represented the last of a northern expansion of organized labor. The problems faced locally required growing levels of intervention from the union's national leaderships, which grew and adopted more complex bureaucratic structures. In Mexico, the industrial unions had pushed, beginning in the 1930s, for national contracts or the takeover of strategic industries. After the nationalization of oil and railways during the presidency of Cardenas, the railway and the oil unions, Sindicato de Trabajadores Petroleros de la República Mexicana and

Sindicato Nacional de Trabajadores Ferrocarrileros, negotiated directly with state-owned firms, PEMEX and Ferrocarriles Nacionales de México. The miners' union pushed the same direction in order to obtain a single national contract, *contrato-ley,* but the government refused to push the multinationals into accepting a single national agreement. This complicated the national organization of the union, although it allowed greater independence of the locals. In response, the union's national leadership started to collect information about firms, wages, rates, technology, and working conditions from the locals. They also compiled and monitored the international prices of metals and the firms' reported profits to produce hundreds of local agreements on the national sphere.[20]

The IUMMSW faced similar issues and had to build an even bigger bureaucratic apparatus. The organization had locals and an executive committee and built several departments that mediated the locals' relationships: health and welfare, political action, administration, and research. The research department was vital, as in its Mexican counterpart, to frame the grievances with individual firms, because they concentrated news about mining firms, collective agreements with other unions, press releases, and relevant correspondence of previous negotiations with the locals. This coordination by the department informed the decisions the locals made, but there were other mechanisms of integration. The locals could organize regional alliances, gather by metals produced, and, more importantly, coordinate for negotiation with the big multinational corporations on a local level. One of the main alliances was among the sections that bargained with the US Company in Utah. In short, the national mining unions were adopting some of the managerial network's structural characteristics inside the corporation.[21]

The generalization of the negotiations between unions and mining corporations was the local expression of the restructuring of industrial relations on a continental level. In the interwar years, organized workers developed a consciousness of their role in corporations' international coordination. The radical miners' unions in North America contested the labor pan-Americanism directed by the states and embraced by the AFL and the CROM. They created alternative coordination based on local alliances on an international level. In May 1936, the general secretary of the SITMRM, Agustín Guzmán, traveled to the United States as a part of the CTM's delegation. Víctor Manuel Villaseñor, the president of the

Society of the Friends of the Soviet Union in Mexico, led the cohort. The delegation refused to have meetings with the AFL leaders and the Pan-American Federation of Labor, which had an alliance with the competing CROM. In contrast, Agustín Guzmán joined John Lewis, leader of the UMWA and the CIO, and established an agreement of communication. Lewis promised to send a delegation to the Congress of the Mexican Miners Union. The IUMMSW, the SITMRM, and the UMWA were soon able to press the multinational corporations into negotiating contracts simultaneously on both sides of the border. In 1940, the corporation had to negotiate at the same time the bargaining for the union shop in Midvale with the International Union of Mine, Mill, and Smelter Workers (Mine and Mill), the strike of the local 50 of the UMWA against the United States Lead Refinery in East Chicago, and the signing of the collective contract #5 with the SITMRM. This time, the union leaders in Mexico threatened with solidarity strikes if the conflicts were not resolved in all bargaining localities.[22]

The beginning of the war led to a harder line of government intervention in the negotiations between the union and the corporations. For the metallic mines controlled by the IUMMSW, contract negotiations during the war had to be arbitrated by the Nonferrous Metals Commission (NMC) and the National War Labor Board (NWLB). This intervention prevented open conflict, but both unions and management kept fighting over the definition of roles inside the corporation. In 1942 and 1943, the NWLB had resolved against the closed shop demanded by the union but in favor of maintaining membership in the different locals of the IUMMSW in Bingham.[23] The Mexican union faced similar restrictions. The mining companies in Mexico had to accept the implementation in 1943 of the Insufficient Wage Emergency Compensation Law. However, the union had to accept the wage freezes for the contracts since 1942. In 1944, the union started a campaign to sign a national contract law for all the country's mining companies. In February, the union resolved a national strike vote for a total of sixty thousand miners, demanding a wage increase of 50 percent and a workers' and owners' convention for signing a national contract law. President Ávila Camacho supported it, but the mining chamber opposed the demand. A national strike started in June, and the sections negotiating with the United States Company in Pachuca, Real del Monte, and in Parral, Chihuahua, followed. Soon, the federal authorities pressed the union into accepting wage increases of only 10 to 12.5 percent. They were not able to reach any further commitment for a conjoint convention.[24]

The war severely limited mobilization capacity, but the boom in industrial metals required for the war gave relevance to a different mineralized area in North America: Grant County in New Mexico. The mining district grew due to the increased demand for lead and zinc, and, as in the case of the tri-state area, in a couple of years, most of the big firms had properties in the area. In 1942 the United States Company acquired a mining property in Bayard, Grant County, transferring a mill from one subsidiary in Fols Road, Arizona, closed as the other gold operations were. Soon after, the firms erected a mining town following the scheme of the barracks and boardinghouses in Bingham and an additional selective flotation plant, as in the Midvale plant. Additionally, the firm acquired a zinc-lead unit in Sonora and increased Parral's operations in Chihuahua. Like other firms, all the concentrates of the lead-zinc properties in Sonora, Chihuahua, and New Mexico were shipped to ASARCO's smelter in El Paso, Texas. This new operational scheme went into full power in the spring of 1943. However, the increase in capital in the area surpassed the combined increase of reserves in Alaska, Mexico, and Utah.[25]

The lead and zinc demand created a new space for integrating miners' different organizational bodies between Mexico and the United States. The IUMMSW created new organizational committees for locals to be able to negotiate on a firm level. At the time of renovations of their contracts, these had annual meetings and published bulletins for all the multinational workers of the Leviathan.[26] The Committee for the United States Company had four locals in the Utah area, #2, #91, #331, and #658, the #444 local in Fairbanks, and the #628 local, a new adherent in Bayard. From the Mexican side, the area was also erupting with labor agitation. In recent years, the authorities in El Paso had repressed some cross-border activism by the communist leaders at Ciudad Juárez on the border. The locals 9 and 11 of the SITMRM in Hidalgo del Parral and Santa Barbara, in US Company and ASARCO properties, were, on their side, extremely active. Both sections participated in the general miners' strike of 1944, generating great local support. This time, the state congress approved a reduction of one day's wage to all the state's officers and employees to support the striking miners.[27]

The creation of a multinational body of workers and the challenge to the form of corporate sovereignty established by the managers was not exclusively the product of agreements among the top leadership of the national unions, nor a simple program of the communist organizations on both sides of the border. The rank-and-file workers had a clear consciousness of their international role in the industrial system in the continent and of the different scales of sovereignty that they were challenging in their organizations. In 1944, the unionized workers of the Mexican division of the United States Company demanded the dismissal and expulsion of Alan Probert, one of the top officials in the division with a large multilocal career inside the corporation. In twenty years of organization, it was the first time that the union included complaints against an official in a strike vote. In his two years there, Probert had gained his fair share of enemies in the plant, the mines, and the workshops. Workers complained that he did not respect their hierarchies, he cursed while talking to them, he refused to replace broken material and safety gear, he did not pay overtime, and, on top, he always threatened to replace them with US skilled workers from other divisions of the company. The technical authority against the workers' organization was revealed in every testimony, and the resistance questioned the national relations that Probert established in the workplace. Paradoxically, the demand for the expulsion of Probert also revealed the insistence of workers on entering the global mining community of the corporation on their own terms: "All these cases, argued one of the union leaders, are humiliations because of the character that [Probert] has. The Director and the Chief of the Labor Department of the company are always courteous and polite, but not this man who sees us very small, and for him, we are nothing but muckers [*mugrosos*]. He is a bad citizen because he is looking for a fight with his nation, while we are looking for the approach among all the united nations. This is social dissolution because he is not following the policy of the good neighbor."[28]

EL PASO, TEXAS

In the spring of 1945, the president secretary of local district number 2 of the IUMMSW, Dan Edwards, started negotiations with the managers of the Big Three in Bingham for the renegotiation of the contract that would end that summer. The union's demands were: seven-hour workdays, forty-hour workweek, bonus for extra hours, two weeks of sick leave, minimum annual wage, wage homologation with the contracts in Coeur d'Alene, and a closed shop. In September, the IUMMSW insisted on the companies operating in the district to form part of joint negotiations. The small operators asked for group negotiations, but Kennecott, Anaconda, and US Company demanded to negotiate separately. As the companies opposed the closed shop and the maintenance of membership accepted by the NWLB in the previous contracts, district 2 transferred the documentation to the national board of the union. From then on, Reid Robinson directly managed the bargaining and tried to reach an agreement with the headquarters of the corporations. The centrality of Salt Lake in the copper and lead smelting industries transformed it into the main stage of conflict between corporations and unions. The union coordinated similar strikes in Montana and Colorado, which followed the events in Utah. This time, the mass miners' organization allowed the union to make the walkout and shut down operations at the mill and the mines, unionizing skilled and unskilled workers. On their side, management was ready to stop operations of the firm for a long time, speculating with the postwar metallic prices. They were also aware that a prolonged conflict could harm the district's radical union's position, as the AFL's crafts unions had some presence on skilled jobs. In other words, all actors were ready to dispute the will of the Automata.

The strike began on January 21, 1946, at 7:00 a.m. in the Kennecott, the US Company, the ASARCO, and the Anaconda Copper operations. This time, the union could entirely stop the companies' operations, having unionized workers in all levels in the US Company and Anaconda and vital workers in Utah Copper. This strength and the normalization of strikes during the New Deal era permitted the IUMMSW and the US Company to agree on the employment of unionized workers for winches and pumps in Lark and US mines, while management kept its watchmen for the duration of the stoppage. On the part of the company, centralization allowed them to operate a single smelter, and they had developed the capacity to turn it off at a relatively low cost. These developments allowed the company to respond to market variations.

From January to June, negotiations with the officers of the companies were managed by Dan Edwards, president secretary for the district. The national strategy of the union allowed the workers to begin to undermine the capacity of the firms for a long conflict. By the end of March, they had reached an agreement with the Anaconda Copper Company, as they had also organized a strike at their properties in Montana, but the other three large firms rejected similar concessions: their new productive structure increased the cost of the strike for the union. After four months, the Kennecott and US Company workers applied to the NWLB for unemployment benefits, but the board rejected their application.

Nevertheless, as the government raised war restrictions, the upward movement of prices in the second half of the year pressured the companies into accepting labor department arbitration. The mediation took place between headquarters and the national board of the union in San Francisco and Washington. Officials of the CIO and ASARCO reached an agreement in Washington on June 11, 1946, with the director of national conciliation Edgar L. Warren's services. At the same time, Robinson managed to secure similar concessions to those obtained from Anaconda with the US Company and Kennecott. Moreover, the end of Utah strikes gave new air to the whole organization. It strengthened the position of the local 444 of Alaska, and the union signed a new contract with the US Company in the next days.

The national organization of the unions in the United States was key in their strategy to combine negotiations in many places at a time. Its coordination with the Mexican workers also resulted in international pressure on the multinational firms. In the summer of 1945, the SITMRM collected information about the different products and price estimations for the following years. They sketched a demand for a 50 percent increase in the silver sector. On September 22, 1945, the Office of Price Administration of the United States increased the price of Mexican silver from 0.45 to 0.711 per ounce troy. Two days later, President Ávila Camacho reestablished the silver appraisal and additional taxes on gold of over 100 percent of the value. In that context, the SITMRM officials tried to implement a joint campaign with owners to negotiate a national contract emulating the conjoint silver conference of Denver, Colorado, earlier that year.[29] The substitute general secretary, Sigfredo Gallardo, started a campaign in which union authorities would support tax reductions for the companies and guarantee further wage increases.

At the beginning of October, the union blamed the secretary of the treasury, Eduardo Juárez, for absorbing over half of the silver price increase. He took "the lion's share, creating a rich Treasury and State and poor industry and workers."[30] Gallardo's declarations sparked a huge controversy during the first days of October. The management of the companies said that the objective of the new taxes was to stop wage increases for workers. They also denounced small benefits and investments in the sector during the years before the war.

Gallardo was leading the negotiations as the general secretary of the SITMRM, while Juan Manuel Elizondo and other members of the executive committee were in Paris during the founding convention of the World Federation of Trade Unions (WFTU). The WFTU was an effort of all anti-fascists, centrist, leftist, and radical organizations that collaborated during the war and supported the Allied countries' leftist unions. Promoted by British and French labor organizations, the WFTU officials in the American continent were the opposition to the pan-American unionism of the AFL. Early in 1945, the AFL had opposed giving visas to some union leaders who participated, along with the CIO, in the federation's organization. In particular, the AFL tried to limit the influence in the continent of Vicente Lombardo Toledano, a Mexican socialist leader and president of the Confederación de Trabajadores de América Latina.

Certainly, the main adversaries of the Pan-American Federation of Labor, aligned with the AFL, were sharing the table in Paris. One close ally of Lombardo Toledano, Antonio García Moreno, was part of the Mexican miners' union's delegation. A native of Nuevo Leon, he was the head of the Partido Popular, a political party founded by *lombardistas,* and had growing popularity among miners. Reid Robinson was also present at the event, along with six vice presidents of the CIO and the secretary treasurer, James B. Carey. During the convention, the Mexican delegation was appointed to organize the Inter American Conference of Union from the Mining, Metallurgical, and Mechanic Industries in Mexico City. The American and Mexican leaders parted ways at the beginning of October, as Robinson and the rest of the CIO delegation left to visit Moscow, part of a goodwill visit to the All-Union Central Council of Trade Unions of the USSR.[31]

When the Mexican delegation came back from Paris, the strategy of the Pachuca strike changed radically. On October 25 union authorities and the vice secretary of labor, Manuel R. Palacios, held a meeting asking the executive's support for a 50 percent wage increase. Government representatives promised to back the wage increase demands but not the tax reductions, so the coordination of owners and workers was immediately broken. Ex-president General Abelardo L. Rodríguez interceded with Ávila Camacho in favor of the owners, and the corporations offered a 27 percent increase in wages if tax reductions were approved. In November 1945, the union addressed wage increase demands with the threat of strikes in all the companies with interests in extraction and processing silver, which would involve some thirty-five thousand workers in a total of thirty-five companies. The companies defended the old collective contracts and maintained the position that the new taxes would absorb most of the price increase. The conflict impacted one of the country's leading industries, as gold and silver production comprised around 40 percent of mining's national value, remained the country's most valuable export, and were a fundamental component in the state's finances. The heart of the conflict was the district in Pachuca between the United States Company's division, the largest silver producer and leader of the Mexican mining chamber, and sections 2 and 146 of the SITMRM, the largest in the country with over five thousand members. The general silver strike started at midnight on January 1, 1946, demanding a 40 percent wage increase, but they were not the only ones revolting. The layoff was twofold: the union against the companies and the companies against the government.[32]

Juan Manuel Elizondo, general secretary of the SITMRM, claimed that the union was in condition to bear the strike until June. By then, they would be capable of calling a new general mining strike and solidarity strikes with more radical industrial unions in the country, the railway, oil, streetcar, power, and typographers' unions. Furthermore, on February 5, the CTM threatened to start solidarity strikes. Simultaneously, the mining chamber secured the support of the Trade Chambers Confederation in their tax relief demands. With pressure increasing, President Ávila Camacho intervened directly, instructing the secretary of labor, Francisco Trujillo Gurría, to reach an immediate resolution, starting in the heart of the conflict: Pachuca. The United States Company and the SITMRM signed collective contract #9 on February 7, 1946, consisting of an increase of 1.25 pesos in all wages, retroactively, and expenses for the union for a total concession of 8 million pesos payable to the SITMRM. In the following days, the rest of the large companies, ASARCO, Fresnillo, and Peñoles, reached similar agreements.[33]

This new space of labor organization across national borders pushed together the miners' national organization structure, re-creating the trust of multinational corporations. The leaders of the SITMRM and the IUMMSW signed an agreement, in El Paso, Texas, in December 1945, in order to coordinate the strikes next year. The pact recognized "a community of problems and interests among the American and Mexican Miners and metal workers." It established several mechanisms that extended the bureaucratic and information structures of the national unions. The organizations agreed to name a permanent cooperation body composed of a delegate of each union, which would have meetings at least once a year in Chicago and Mexico City. The organizations decided to send representatives to each other's conventions, and share contracts, legal dispositions, internal working rules, wages sheets, organization statues, production indexes, life costs, productivity projections, ore expectations, social welfare systems, accident reports, and health statistics, among other relevant information. Finally, the unions committed to giving material, political, and moral relief for the other unions' strikes and refusing to treat metals produced by strikebreakers on the other side of the border.[34] The agreement's last objective was to extend the binational alliance "con los demás paises de America." The agreement was signed in both English and Spanish by the leaders and ratified by their respective conventions in January 1946.

"The Western hemisphere at this time seems to be pretty much lined up in the struggle of labor and capital," declared Reid Robinson in his inaugural speech at the 1946 convention of the International Mine and Mill Union. Two years before, the convention had as its objective defeating fascism, under the slogan of "full production for final victory."[35] He regretted that this alliance with sovereign powers in the war did not transform into long-term democratic planning after the war. He was, nonetheless, confident that the democratic and centralized structure of the union, and its integration into the WFTU, would guarantee their growth in the coming years. He was wrong. Robinson resigned a few months later because of the increasing pressure of the CIO's anticommunist sectors; John Clark replaced him and established a joint leadership with Treasurer Maurice Travis. In 1947, Congress approved the Taft-Hartley Act, which made demands for closed shops illegal and forced union officials to swear not to be communists. The next year the CIO expelled the IUMMSW. From then on, the UMWA and the IUMMSW engaged in an anti–Taft-Hartley campaign and competed with the AFL and the CIO for the locals in the metallurgical and mining sectors. The organized body of workers, who had been able to coordinate actions across several corporate, national, and empire sovereignties, was having trouble maintaining a single identity.

BAYARD, NEW MEXICO

In April 1947, Virginia and Clinton Jencks arrived at the mining camps in Bayard, New Mexico. They came from Denver, where they had lived for some years, and came to the booming area as organizers of the IUMMSW. When they started scouting the area, they found that the long-standing tradition of job and urban segregation was still real. Mexican and Mexican American workers could not access skilled jobs; they had separate payrolls, separate changing rooms, and a dual wage system. The boarding and urban arrangements of brown workers remained separate from their white counterparts. The companies negotiated rates with the white workers affiliated with the AFL, so the Mexican and Mexican Americans increasingly organized under the CIO's flag. Nevertheless, racial segregation remained an obstacle to a common organization, but the mining operations' geographical dispersion made it difficult to integrate unionized workers. The most important operations were the Vanadium mine of ASARCO, four mines of the United States Company, and the Empire Zinc mine, of the New Jersey Zinc Company. In the next months, the Jenckses reorganized the labor agitators in the different companies and formed the local 890 of the IUMMSW. Eight of the ten organizers that promoted the local were Mexican Americans, and three of them, Ishmael Moreno, Ray Leon, and Charles Morell, worked as miners at the US Company.[36]

Mexican Americans made up 90 percent of the local 890 and soon started to oppose some of the firms, which prompted the firms to respond. In 1948 Kennecott, the US Company, and ASARCO refused to bargain with Grant County contracts if the union leaders did not sign the non-communist affidavits required by the Taft-Hartley Act. Jencks rebutted successfully that year, as the union sustained that the affidavits were not a requirement for bargaining and only a requirement if they asked for NLRB intervention. The defense of their bargaining position worked, at least temporarily. After the death of three workers in the Bullfrog and Continental mines of the US Company, the menace of a wildcat strike allowed the local 890 to obtain the Bureau of Mines' attention and bargain the participation of the union in the safety committees.[37]

Nonetheless, the camps' conditions were still tough for the Mexican workers, as workers' bodies had to deal not only with trauma inside the mines but also with industrial poisons outside. In February 1949, the laboratory at Midvale analyzed water samples from the mining camps' domestic water. Through internal research of the US Company, the general chemist of the plant, C. A. Greenwood, found that the mining camps' domestic water surpassed the Public Health Tolerance Standards in the content of arsenide, lead, copper, iron, zinc, manganese, and sulfuric oxides. The head of the agricultural department concluded that the water was "unfit for human consumption without prior treatment."[38]

Things on the other side of the border were getting difficult for the miners in the SITMRM. Radical industrial unions linked to the Communist Party, from the telephone, power, aviation, cement, streetcars, oil, and mining industries, under the railway union's leadership, abandoned the CTM Congress in 1947 and formed the Central Única de Trabajadores (CUT). The government quickly began to attack the new organization. In October 1948 federal authorities, using police and the military, imposed Jesús Díaz de León, El Charro, as the railway workers' general secretary. El charrazo was a policy during the next years that consisted of purging leaders in antagonistic unions. Internationalism was the mark of communism in Mexico, and the combination of xenophobia and anti-communism in the US Southwest condensed ideological and racial prosecution against brown labor organizers. Workers' organizations required, in that context, some other fronts in defense of their civil liberties. In February 1949 in Phoenix, fifty delegates of the sugar beet fields of Colorado, Los Angeles factories, a cotton field in Arizona and Texas, and the miners in New Mexico founded the Asociación Nacional México-Americana (ANMA) to promote "the protection of the civil, economic, and political rights of the Mexican people in the United States."[39] Unlike some other Latino civil liberties organization,

ANMA did not require US citizenship for being a member and established locals in several Southwest regions. One of the first actions in El Paso was to facilitate a second meeting between the national leadership of the miners' unions in Mexico and the United States. Between April 21 and May 2 of 1949, the leadership of IUMMSW and SITMRM had a new summit in El Paso to endorse their Pact of Friendship and Mutual Aid. The agreement included a North American integration of the unions in Mexico, the United States, and Canada, and projected to unify other Latin American unions. On the new pact, the unions took a shared demand to stop the requirement of passports for union leaders to cross the borders. The increasing power of the state on the borders was breaking the bridges for the international organization.[40]

The year 1950 was decisive for the international body of workers that had united in El Paso. The ANMA started a peace campaign among Mexican Americans protesting the war in Korea. At the same time, the CIO and AFL battled all organizations among the miners, competing for representation in the different mining camps with the Mine and Mill Union. The expulsion of the IUMMSW from the CIO also empowered management's efforts to destroy the unions or install white unions in their camps. At the beginning of 1950, ASARCO promoted divisions among the adherents of local 890 in their properties. By its side, the campaign of the US Company to install a different union succeeded; the firm installed the Grant County Miners Association at all the properties, replacing the representation of the locals 638 and 890 of the IUMMSW. Jencks and the rest of the organizers were restricted, then, to their organization at the Empire Zinc Company.[41]

On the other side of the border, the union was facing similar tensions. In 1949, part of the miners' union's leadership had formed a new organization, the Union General de Obreros y Campesinos de Mexico (OGOCM), which formally adhered to the WFTU. The membership of the convention, close to the radical *lombardismo,* was estimated at around thirty thousand people, and it tried to hide the communist activism of the leaders as they faced stronger pressure from the government.[42] In May 1950, illegal delegates raided the miners' convention and elected a governmental loyalist, Jesús Carrasco, as general secretary. Meanwhile, the rest of the delegates moved to another facility and elected Antonio García Moreno, the general secretary of the OGOCM, as the new general secretary. Most of the sections were *garciamorenistas,* but the Ministry of Labor recognized Carrasco's faction, denied recognition to the dissident locals, and instructed the local governments to pressure them into accepting the official leader.

The locals of the IUMMSW in New Mexico and the miners of the SITMRM in Coahuila started simultaneous strikes in 1950 against zinc mining companies. On October 16, 4,500 miners of local 14 of the SITMRM walked out of the Mexican Zinc Company, a division of ASARCO located in Nueva Rosita and Cloete, Coahuila, demanding better work conditions and higher wages. The next day, less than six hundred miles away, local 890 of the IUMMSW walked out of the Empire Zinc Company in Bayard, New Mexico, an area that was controlled by the US Company, Anaconda, ASARCO, Phelps Dodge, and the New Jersey Zinc Company. Bayard's strikers demanded the homologation of their salaries to the miners in the rest of the companies in the area, equal pay, and desegregation of the housing between white and Mexican Americans. In October, Antonio García Moreno asked Maurice Travis for help with the Coahuila strike, according to the El Paso agreement. Nonetheless, the delegates of the IUMMSW were, at that moment, present at the convention of the SITMRM led by Carrasco. The treasurer of the IUMMSW, Maurice Travis, traveled to Mexico at the beginning of 1951 to witness the dispute over leadership, while John Clark, the president of the union, maintained relationships with the Carrasco faction. In other words, Travis and the communist leaders of the IUMMSW maintained contact with the miners' union dissidents but avoided an explicit conflict with the leadership of Carrasco. One way to attain this was through the ANMA. The civil rights organization was the last link between the miners' unions in Mexico and the United States, assisting the communist locals in Mexico and promoting Mexican Americans' civil rights in the West.[43]

The zinc labor conflict in Bayard and Coahuila continued for the rest of 1951, exceeding the strike time that the firms and the unions had sustained in 1947. As the dispute involved more and more of the mining community, as in the case of the activism in the tri-state area, the women were more present in the visible leadership of the resistance. Whole families of miners in Santa Rosita organized in the summer a march of over a thousand miles to Mexico City, which was known as "The Hunger Caravan." In Bayard, the women took over the picket lines and increased their centrality in the workers' organization. The demands of the strike were not restricted to the workplace but extended over the living conditions in the camp. Sanitation was one of the main demands to the company, as only the housing for white workers had running water. By December 1951, the Mexican miners pulled off the strike against the Mexican Zinc Company, and the locals supporting García Moreno accepted Carrasco as the general secretary. In January 1952, the strikers at Empire Zinc settled with the company, winning a heartfelt victory over management. Besides eliminating the "Mexican wage" in the labor contract, soon after the company installed hot running water in the housing for all the workers. Still, the radical miners in Mexico and the United States met later that year, during the Interamerican Mining Confederation convention in Mexico City, organized by García Moreno.[44]

BOSTON, MASSACHUSETTS

"How should I begin my story, that has no beginning?" asked the narrator of the protagonist of the film *The Salt of the Earth*. During the fall of 1951, a group of blacklisted film producers, the Independent Productions Corporation Inc., had visited the Empire Zinc strike's picket lines. After the end of the strike, the crew filmed in the same area, and most of the actors were the local labor organizers. As in the short film about the tri-state area by Sheldon Dick years before, the production appeared as a combination of fiction and documentary, as director Henri Bieberman employed miners, including Clinton Jencks, to portray themselves. The arc of consciousness of the mine's materiality in the community's bodies and inside the miners' household was portrayed by a miner's family. The film followed the increased role of Esperanza (Hope) in labor organizing; the changes in her marriage with one of the union leaders, Ramón; and the unveiling of women's roles in and outside of work. Rosaura Revueltas, a Mexican actress from the Revueltas leftist dynasty, played the role of Esperanza. Her casting both showed terror of Hollywood actors in participating in the film and the transnational alliances of the miners at the time.[45] Two months after the film's release, the courts banned the movie, Rosaura Revueltas was deported, and the International Mine and Mill Union faced a severe prosecution of the national leaders.

"My name is Esperanza, Esperanza Quintero." In the years before, the radical leadership had had great hopes of multinational workers' integration, but the possibility was fading away. The closeness of Travis and García Moreno continued in the following years, but soon that informal alliance was also under federal intervention. On October 7, 1952, Maurice Travis appeared before the Subcommittee to Investigate the Administration of the Internal Security Act in Salt Lake City. It was the second session devoted to the Communist Domination of Union Officials in Vital Defense Industry, and the members of the judiciary committee had concluded that it was Travis, not President John Clark, who ruled the union. The interrogator started listing Travis's different residences and then asked the number of affiliates in Alaska, still a territory, and Canada, trying to establish the union's international character. The critical questions, nonetheless, were about the connections down south.

Dr. J. B. Matthews, former research director for the House Un-American Activities Committee (HUAC), identified the American Continental Congress for World Peace, held in Mexico City in 1949, as one of the central communist enterprises in the Western Hemisphere. Harvey Matusow, an FBI informant, had infiltrated the organizations by approaching Clinton Jencks and was vital on the HUAC trials, combining the racist rejection of Mexican politics and the era's anti-communist scare. Matusow declared that the delegate to the United Nations' Human Rights Conference, George Stary, asked him to provide biographic documents to the Mexican embassy in New York. They would, he claimed, process them to involve him in a mission to collect atomic secrets from the Sandia base in Albuquerque. Matusow alleged that the Asociación Nacional México-Americana (ANMA) was a communist front organization set up to use the plight of some of the Mexican Americans as a way of getting communist propaganda and communist work across the border. The script of the trials shows the very simple combination of anti-communism, racism, and xenophobia that defined the HUAC era. Matusow declared that the orders for the copper strike during the Korean War had come through the southern border, as union leaders from the IUMMSW established a network with the miners' union in Mexico, through the Communist Party, and allowed the intrusion of Soviet elements into the country: "People would come in, Communists, illegal immigrants, you might say, Soviet agents, and so forth, would come across the border at various crossings, such as El Paso, coming over from Juarez." The next month Clinton Jencks was arrested by the FBI, and one year later, Matusow described, in his book *False Witness,* how he had been paid to infiltrate the union and then lie in federal trials by the HUAC.[46]

The trials marked a breaking point in the trajectory of collective agreements and union organizations in the hemisphere. From a system of individual contracts at the beginning of the century, the workers had organized locally, nationally, and then transnationally into a highly complex body of workers. These new unions were able to dispute with global corporate powers and negotiate with national states. Nonetheless, a new international alignment of corporate and state sovereignties was able to restrict their capabilities in the following years by intervening in their leaderships.

A new legal culture had also developed in the labor legal scholars at the time. In the years after the NLRB trial, the US Company's old legal team quickly became relevant in the national arena, but they all maintained a Boston residency and started teaching at Harvard Law School. Henry C. Moses focused on private practice, became legal counsel of Mobil Oil Company, and was appointed vice president of the multinational consortium years later. Charles Wyzanski became a federal judge in 1941 and was appointed a member of the Harvard University Board of Overseers, a position that he occupied for the next thirty years. During the war, Archibald Cox became part of the assistant solicitor's office in the labor department, and between 1943 and 1947, he was part of the Wage Adjustment Board in the National War Labor Board. After the war, he became a full professor at Harvard Law School, where Moses and Wyzanski continued teaching, and, along with Wyzanski and John T. Dunlop, of the economics department, became one of the leading exponents of industrial pluralism in Cambridge.[47]

After the war, the industrial pluralists reflected on the nature of collective bargaining and its legal place in a democratic society. Wyzanski and Cox, as most of the proponents of that kind of liberalism did, framed the existence of collective bargaining as a stabilization tool of industrial relations. The association of workers was just the natural counterpart of the association of owners and managers, and those actors had no intrinsic contradictions; they were collective bodies forming contracts over them. As the collective agreement was an extension of the private form of contracts, as union officials and management were trustees of labor and capital, collective workers had to obey contracts until their validity date.[48] Based on the New Deal Wagner Act, Wyzanski and Cox tried to rationalize the role of the state in the resolution of the conflicts among the actors.[49] Wyzanksi signaled, in 1942, that the compulsory arbitration of labor conflicts during the war created a disputed space of sovereign rule inside the corporation. A resolutive board would eventually meet defiance from a strong employer or union leader, "and that defiance will gain strength because the Board cannot point to the breach of an established rule."[50] For Cox, the role of the arbitrators in labor disputes was to look for breaches in the contractual obligations of the parties, and not in the definition of general grievances between labor and management. The clauses of arbitration in the collective agreements were, in his words, areas of "joint sovereignty."[51]

Nonetheless, Cox and Wyzanksi considered that the anti-communist regulation of labor organizations was necessary. The Taft-Hartley Act established that the National Labor Relations Board would not intervene in a collective bargaining dispute if the union officers did not have an affidavit of not being part of the Communist Party. Cox admitted that the new regulation would increase unions' difficulty in being recognized as bargaining agents, favored the trade against the industrial organizations, and was arguably unconstitutional.[52] Nevertheless, Cox argued that union was not immune to government regulation and that there was "a recognizable public interest in the prohibition of wasteful strikes and picketing."[53] This freedom of the unions and the body of organized workers was nonetheless bounded inside the democratic society, as the firm's space of sovereignty was a common ground for all the American corporations. For Judge Wyzanski, communism was one of the examples that restricted the right of association, where the state could establish an exception to the rule of law to preserve its sovereign integrity. "We lawyers know that in private matters, an agent cannot act for two principals on subjects where their interest is antagonistic. So in public matters, a man cannot serve two nations in a field where their interests are opposed."[54] These regulations on the subjectivity of the workers, and the space of sovereignty in which they intervened, were central legal strategies for redefining the obligations of the new legal collective bodies of workers.

In the next years, Cox became close with the young senator John F. Kennedy. By his side, Judge Wyzanski participated in the discussion for the development of the Ghanian constitution and started collaborating with the Ford Foundation in education plans to expand American influence in Africa against Chinese and Russian communist influence in the area. He noted that Ghanian unions were more similar to Soviet unions than to US unions and that the collaboration of USSR technicians made them very popular among the educated and popular classes. "Everything suggests that the USSR is placing in the Congo technicians whose role may someday be to stage a Communist coup," he noted, and advised a greater exchange of American private universities to elite bureaucrats in those regions, in particular doctors. He warned, nonetheless, that those exchanges should be to Africa and not the other way around, as the racial segregation in the United States might generate resentment from the black elites coming back home. "Because of U.S. prejudice against black people, medical, business, and other education probably ought not to be too heavily transferred to the U.S.,"[55] he concluded.

The physical body of workers in those places was at the center of labor agitation, and the industrial physicians stayed in the focus of anticommunist surveillance. Alice Hamilton was under growing scrutiny from the FBI regarding her political participation. All the organizations in which the Hamiltons participated—the American Friends of Spanish Democracy, National Committee for People's Rights, American Committee for Democracy and Intellectual Freedom, Chicago Civil Liberties Committee, Women's International Congress, the Civil Rights Congress, Committee for Peaceful Alternatives, and the American Committee for the Protection of the Foreign Born—were considered front organizations of the Communist Party.[56]

In those years, the state and corporate sovereignties realigned against the imaginary menace of communist workers. The organized body of workers had lost its capacity to extend beyond national borders in their quest to match the corporate Leviathan. Their conquest of the international corporate space, both working and living, had been successfully stopped by the combination of corporate and state sovereignty. The expansion of the frontier of the Underground Leviathan was contested but remained unaltered for the most part. The limits of the corporate space had grown from the workplace to the living areas and from the company houses to the worker's lungs, but they had also expanded beyond, into the rivers and winds of the valleys, and into the cattle and crops of the peasants. The sovereignty of the Leviathan did not end on its territorial dominions, their formal properties, nor on its leading actors, the subjects and rulers of the corporation, but extended into the general community, creating a liminal space of power: a space of exception.

CHAPTER SIX

Space of Exception

God created iron but not that which is to be made of it.
He enjoined fire, and Vulcan, who is the Lord of fire,
to do the rest. From this, it follows that iron
must be cleansed of its dross before it can be forged.
This process is alchemy; its founder is the smith Vulcan.

—Paracelsus

AN UNDERGROUND RIVER

There was always a volcano under their feet. Magma charged with water, acids, and metals rose from the interior of the earth. As it advanced to the top, it pushed rocks, forming mountains; it broke part of the crust and intruded itself into other, softer stones. If the pressure was enough, magma came out in an explosion, making a hole in the earth, like a balloon. The materials expelled formed various landscapes as they cooled in the exterior, creating a crust from the active fluids underneath, a complex combination of all the elements found in the planet. Other times, the lava did not open the skin in the ground, and the immense force petrified underground. These intrusive and igneous rocks, formed under earth's surface, changed some other, older structures—the sedimentary rocks—creating faults in them, forming mountains, cliffs, canyons, and valleys.

Metals were one of the toughest components in the mixture coming from the center of the earth. As magma cooled, metals often concentrated in deposits combined with sulfur. Each configuration of the ions in atoms produces different geometries. Lead sulfides, such as galena, formed opaque octahedral rocks; pyrite, containing iron, and argentite, containing silver, formed cubic sulfides; sphalerite, containing zinc, and chalcopyrite, the principal ore of copper, form tetrahedral structures; porphyry, containing copper and molybdenum, cooled in two phases, combining with some other magma into a low-grade ore of a round texture. Molten magma additionally forms quartz, a crystalline mineral composed of silicon and oxygen also on a tetrahedral structure. Quartz is the second most common mineral formed in the crust and, when metal sulfides form on a vein, it precipitates and crystallizes around them. Other crystals also form around the intrusive metals—calcite, magnesite, or dolomites—just slightly softer than the metals themselves.

Sometimes, the material on the earth's surface was exposed to erosion and watering, breaking the minerals into sand and gravel. When settled into water bodies, the different layers of the ground compacted into sedimentary rocks, limestone, sandstones, mudrocks, or coal. The eroded material that does not transform into sedimentary rock forms the superficial landscape, ground, gravel, clay, silt, and sand lying on top of the rocks. Transported by water, igneous and sedimentary rocks can form deposits again, as the different weights of minerals allow the gravitational concentration of ores. Gold nuggets, sands with a high concentration of precious metals, are formed by this phenomenon in consolidated and nonconsolidated deposits. This process, for instance, formed the alluvial sands with a high concentration of gold in Alaska.

Other times, the metals remained hidden in the landscape, covered by sedimentary rocks, soil, forests, and living beings. In Bingham, Utah, sulfides of copper and lead cut the sedimentary rocks, sandstones, and limestones, forming the typical sharp edges of the Oquirrh range. In Hidalgo, Mexico, the intrusion of metal sulfides deformed the andesites, which are softer volcanic extrusive rocks, creating the sharp curves of the Sierra Madre hills. Below those landscapes appeared a secret geometry of rocks, broken minerals, and underground rivers. The static magma looked like roots from the bottom of the earth, hard rock veins in the mountains covered in multicolor crystals.[1]

The United States Company was part of an underground stream that communicated the dusty valley of Salt Lake to the electrical wires in Pittsburgh, and the financial markets of New York City to the dry cliffs of Real del Monte. The corporation was a very complicated machine with a central mission: to transform the metal sulfides found in the rocks into refined metal used in industry or as currency. The company had to extract the ores from the mines, transport them and crush them, and clean them from the quartz gravel and rocks. This primary process was standard for all the company's subsidiaries, but the transformation of sulfides of different metals required a different process. The lead and copper ores of Utah oxidized through combustion in blast furnaces. In the lower part of the furnaces, big pumps injected significant air volumes, allowing the combustion of the fuel on top, coke from coal. Sulfur and metals separated in the high temperatures and then associated with the free atoms of oxygen. Copper and lead oxides dropped to the bottom, while the stack expelled sulfur oxides. Metallic oxides were then refined into bullions and sent to the West. Sulfur dioxide and trioxide escaped to the atmosphere, but they could also stay in the surroundings. In either case, when combined with water, whether in the clouds or with the surface dew, it transformed into sulfuric acid.[2]

Cyanidation was the most effective method for silver extraction from sulfides. A solution of potassium or calcium cyanide with water was combined with the crushed ores. The suspension was then pumped into agitation tanks that allowed the insertion of oxygen into the mixture, dissociating the sulfides and the cyanides. Sulfur forms potassium, sodium, or calcium sulfates. Zinc dust was used to replace the newly formed cyanides' silver and gold, precipitating the more precious metals. Metallic lixiviates were later refined in reverberatory furnaces, producing bullion. The sulfates, gravel, slag, and residues of zinc cyanide were disposed of using water currents. The combination of some heavy metals, crushed ore, and cyanide was highly toxic.[3]

Matter did not disappear after it left the company plant, and the firm had to carry out the expulsion of those materials. Refining and wasting were two sides of the same process, and both implied the definition of limits in the operation of the firm. Designing those boundaries implied barriers between internal and external actors, processes, and properties. In other words, the firm was a sovereign that had to constantly invent the limits of its kingdom.

In previous chapters, we analyzed the continuous territorial and visible expansion of the firm in the territories by surveying, identifying, acquiring rights to mineral bodies, extracting them, processing them, tracking them, and selling them. The sovereignty of the firm over the objects of work appeared, up until now, as a legal and geological definition of mineral and as a form of the minerals deposited in the landscape. This territoriality allowed the material existence of the several bodies of the Leviathan, but another body stayed over the formal boundaries of the firm: a body of waste. This body, extraterritorial in many ways, established invisible limits of the firm over the landscape. As navigator of underground metal rivers, the ghostly existence of the firm also had an impalpable toxic existence on the surface.

This chapter argues that the difference between the firm's formal territorial boundaries and the informal and invisible limits of its body of waste generated a *space of exception*. Hobbes argued that the actions of the sovereign could not be accused of being unjust by the subjects, as they represented the collective will.[4] Following that concept, Carl Schmitt considered the *exception* as the defining characteristic of the sovereign. The *exception* is the sovereign's suspension of law, the decision over the limits of absolute power beyond the common law.[5] The concept of exception is the foundation of this definition of sovereignty, the exception being a "zone of indistinction between outside and inside, exclusion and inclusion," but it is usually thought of as a constraint period. Analyzed as a *state*, the exception appears only as an emergency. Even if the ability to declare it is a sovereign power, the exception is constrained temporally by political decisions. The *state of exception* is an event, a moment when the constitutional norms do not apply, an interregnum when the legal order resides on decisions and not on norms.[6]

In contrast, analyzed as *space*, the exception is a territorial realm defined by material boundaries, invisible limits, that are not constrained by time. It is a space where the law of the land does not apply, a place outside the *lex terrae*, but not outside the sovereign's power.[7] This space of exception is, at the same time, extraterritorial to the firm, in so much as it is located outside its legal limits, and territorial to it, as a zone where the corporation ruled. This chapter will show that these limits changed over time as the corporate body expanded its rule, but they were always determined by the strategic and political decisions of management. The existence of this space of exception was not a mere failure of the property rights of the firm, a simple externality on Coasian terms, but it was constitutional to corporate sovereignty.[8] This new form of ghostly existence of the US Company was revealed in the conflicts over land in the Hidalgo region and the Salt Lake Valley.

THIRD DISTRICT COURT, SALT LAKE

In the Book of Mormon, the prophet Lehi, an Israelite from the tribe of Joseph, had six sons: Laman, Lemuel, Sam, Nephi, Jacob, and Joseph. The sons of Laman failed to follow the Lord's commands, who had vested authority over Nephi to be their ruler and teacher. As punishment, "Lord God did cause a skin of blackness to come upon them"; their seed was cursed, and they became idle and sought for wilderness. They dispersed, according to the gospel, in the Americas and the Pacific Islands. Despite their betrayal, the Lamanites, as descendants of Lehi, could reclaim the blessings of Abraham and were virtuous when Jesus reached the American continent.[9]

Mormon colonizers believed that the Indian tribes in the Americas to be Lamanites. This shaped their attitude toward them: separation and negotiation. The mixture was strictly forbidden, but no fights exploded in the first years of the pilgrimage to the West. As the journey was prolonged, the policy of the church hardened. When the group led by Brigham Young reached the Salt Lake Valley, tribes of Utes and Shoshones approached to trade cattle and munitions. After a while, some Shoshones started to demand payment for the use of the land, whether in munitions or in corn. Church authorities opposed any payment, denying any claim from the indigenous peoples over the territory. "No man shall have the power to sell his inheritance for he cannot remove it; it belongs to the Lord," said Heber C. Kimball of the Quorum of the Twelve Apostles of the Church of Jesus Christ of Latter-day Saints.[10]

The Mormon agrarian ideals of self-sufficiency, independence, and planning had been reinforced in the previous years of persecution in New York, North Carolina, Wisconsin, Illinois, and Nebraska. The church distributed the plots in the promised land and planned Salt Lake City—the temple, the roads, and the stores. The settlers installed irrigation systems fed by the creeks in the north of the valley and constructed mills in the river that separated the basin from the mountains in the West, which they called Jordan. In their collective memory, they were transforming the desert into the richest land of the earth through work, removing the curse on the land.[11] Soon, the tribes populating the Great Basin were considered a nuisance for the accomplishment of that agricultural dream. In 1850, Brigham Young initiated the first campaign against the Utes, with the support of the Mormon militia and the US Army Topographical Engineers. In the next few days, twenty warriors with their families surrendered to the settlers. The expedition murdered

all of them, including women and children, and the army surgeon decap-
itated the bodies, for research that he considered scientific. "I am sent
now to confiscate all their property—and then put them in the heat of
battle and kill them," said Brigham Young after one year of an exter-
mination campaign.[12]

As the agricultural frontier expanded, the Mormon farmers and
ranchers expelled the Utes, Paiutes, and Shoshones. Indian chiefs continually sought mediation with the religious authorities, but any resistance by the part of the natives, usually expressed as small attacks or
cattle stealing, was quickly followed by slaughter campaigns. In those
years, the Mormons defended the promised land against another threat:
the white rushers in the mountains. Brigham Young, concerned with
the migration of gentile rushers, forbade the exploitation of the mines.
In the 1850s and 1860s, federal troops installed a base in the Wasatch
Mountains, a high point of surveillance over the Salt Lake Valley. The
troops participated in the violence against the Indian tribes but also
clashed with the Mormon militias. After a while, the soldiers found the
metallic ores of the range and started to exploit them in small mines and
mills. For the rest of the century, the Mormon agriculture dealt with a
double frontier: against the free tribes of the basin and the white gentiles of the mountain.[13]

Benjamin Winchester was one of the first Mormon farmers to settle
in the Salt Lake Valley, obtaining twenty acres of land next to the Jordan
River. In the 1860s, with the arrival of the railroad and the growth of
the mining activities, some farmers started to work in the small companies in Bingham and Park City during the cold months. He never
did. He wrote to his son in 1866, in an especially hard winter: "I would
never think in going to the mines, as almost everybody that went from
here came back sick."[14]

Nonetheless, the Winchester family could not resist for long some
form of integration into the boosting activities of the metal business.
One decade later Stephen, son of Benjamin Winchester, started to work
every winter in one of the small metal furnaces that operated in the
Jordan River. The small plants treated the ores from Bingham Canyon
and then transported them to the railroad station, in the north of the
city. In 1901 two new companies built modern blast furnaces in West
Jordan, as the prices of copper and lead increased. ASARCO installed a

plant in Murray that specialized in treating the ores of the Utah Copper Company, also a Guggenheim company. In West Jordan, the United States Company built a concentration and roasting plant for processing the ores from the United States Mining Company, another firm of the holding. Some ten years after its installation, the US Company plant, now called Midvale, treated alone ten times more ore than did all the small furnaces in the valley in 1871.[15]

This transformation did not go unnoticed in the neighboring farms. On December 3, 1903, S. M. Whitmore filed a civil complaint against the United States Smelting Company and the Bingham Copper and Gold Mining Company, in the third district court of Salt Lake. Whitmore was a farmer who possessed land half a mile from the smelter of the US Company in Midvale, where he grew hay, vegetables, orchards, corn, and wheat. The farmer stated that the smoke expelled by the blast furnaces of the plant, combined with the moisture of air and the dew, produced acidity that affected his lands. In days with enough wind, dense fumes extended for over a mile on Whitmore's lands. "Prior to the erection of the smelter of said United States Smelting company, this plaintiff was not aware of the damaging character of the smoke, fumes, and dust as herein set for and this plaintiff alleges that the fumes and dusts discharged from the smelter of the Bingham Copper and Gold Mining Company, while damaging, and injurious to the premises of the plaintiff and other nearby and adjoining lands, do not contain nearly as large a portion of substances poisonous or potentially poisonous to the soil and vegetation as the smoke and fumes from the smelter of the said United States Smelting Company."

On that year, apart from the loss of several trees and crops, twenty-two cows and five horses got sick and died. The case of Whitmore, the first made against the smelter in the Salt Lake Valley, was prototypical of the legal processes of the farmers in the area in the future. Beginning with a rough knowledge of the damages, starting from the visible smoke and dust, he was forced to research the reactions of sulfur dioxide emissions with water. On the other hand, in the beginning, he only demanded the repair of damages for $1,965. Once convinced of the permanent nature of the pollution, he demanded the end of the company's operation or the forced sale of his lands, for $15,000.[16]

Between 1904 and 1905, the United States Company faced eighteen similar cases, on a total of $66,000 in damages. Besides the destruction of crops, the farmers started to claim personal damages for the flue dust. Black powder was deposited on the houses of West Jordan, darkening the walls, floors, beds, and furniture, and affecting the health of the residents. By the end of 1904, the rest of the farmers of the valley, James Godfrey and four hundred other landowners, started a legal suit against the Bingham Copper and Gold Mining Company, Utah Consolidated Mining and Smelting Company, ASARCO, and the US Company.[17]

In November 1906, the court decided favorably for the farmers. The case *Godfrey vs.* ASARCO *et al.* established legal precedent over property rights, as it stated that the fumes were a collective nuisance, not solvable by individual settlements. Fumes and smoke from different sources combined in the air, and the winds affected the properties differently. Compensations, said the court, forced contracts between the parts and assigned prices over nonquantifiable goods. Consequently, the court resolved that all the four smelters must stop their operations, and two of them, the smaller ones, ended up shutting down permanently. ASARCO and the US Company argued that the district would be hugely affected, as they employed thousands of workers in the valley and thousands more in the mines. While the court refused to take the economic losses as an argument, it pushed for an arrangement with the farmers. The firms agreed to pay indemnifications to the farmers for two years, while they built baghouses to filter the fumes and capture the dust of the smoke. Additionally, they would increase the height of their stack, in order to avoid the surface contact with the neighbors. With the new technology, the corporations stated, they would not expel gases with a concentration of sulfuric acid over 0.75 percent. This discharge should not damage the crops or cattle, nor generate any nuisance to the neighbors.[18]

The case required a profound transformation of the company officers and the redefinition of the boundaries of the corporation. In 1907, the firm employed Clarence B. Sprague, a chemist educated at the University of Utah who had designed a method of substantially reducing the solid particles in the fumes. Sprague's method treated the smoke with zinc oxides and calcium, precipitating the sulfuric acid particles into zinc sulfides and water. After that, the fumes would circulate into a circuit of wool bags that collected any flue dust remaining. In 1909 the new baghouse started to treat the gases from the Midvale plant, and

Sprague reported that the gases were still highly destructive, corroding the bags even after treatment. The managers of the filtration system had at least one fire every week, and most workers resigned the first day they had to work in the dusty environment. Still, the company increased their operations after 1909 and extended the method to their smelters in Kennett, California, and the lead refining plant in East Chicago. Two plants of ASARCO adopted the Sprague process, in Murray, Utah, and Perth Amboy, New Jersey, and the US Company obtained patents in Mexico, France, Germany, and England.[19]

The farmers won in court, and the company had to modify their operations, but the damage extended to the valley. The frontier established by the Mormon settlers against the smelters could not stop the invisible expansion of the company in the territory. In the following years after the construction of the baghouse, Sprague continued his investigations on the harm of the gases on the vegetation in Utah and California. He discovered that sulfuric oxides were highly destructive for the leaves, even in small quantities. Moreover, the baghouse could only stop the solid particles, but not the gases liberated into the atmosphere. It could not transform the sulfuric oxides into sulfuric acid in a controlled environment. In short, the baghouse clarified the smoke, removing the solids, but the invisible carbon and sulfuric oxides expanded the invisible limits of the company operation.[20]

PACHUCA, HIDALGO

Repartimiento was the colonial system that forced indigenous communities to work in the Spanish industries. A Spanish lord would ask the Crown for a certain number of workers from the neighboring communities. Communities on the República de Indios could negotiate the number of workers sent to the mines; the laborers had to receive a wage in silver, they would rotate every week, and the companies had to guarantee minimal living conditions. At least six hundred *repartimiento* Indians worked in the mines of Pachuca in 1580, but that number quickly fell after the demographic catastrophe of the coming years. One century later, only fifty-four Indians labored the mines of the district, and the companies started to ask the Crown for workers from farther away, the northern and east Sierras.[21]

In the middle of the eighteenth century, the mining companies did not respect those regulations of forced labor. In towns, they recruited more Indians than they ought and kidnapped indigenous merchants who sold their products in the mining towns. The repartimiento laborers came out of the mines practically dead after being forced to work twenty-four hours straight or died by poisoning in the treatment plants with mercury. The pueblos rebelled against the system. They started legal processes, delayed the delivery of their quotas, or denied the provision of workers altogether.[22] In April 1757, the last group of the repartimiento workers of the pueblo of Atocpan, an Otomí community, rebelled and escaped from the work on the mines. After some days, the rebellious workers were over one thousand. They had been forced to work in the mines of Real del Monte, but rebelled, hid in the mountains, and then returned to their villages. In the next days, the indigenous authorities had a meeting with the officials of the Crown and the religious hierarchy, rejecting new recruitment for meeting the demands of forced labor. They would not force "their children to work in the mines; they would rather abandon their villages and desert the jurisdiction."[23]

The mines that required so much forced labor were the property of Pedro Romero de Terreros. In 1739, the company received authorization from the Spanish crown to exploit the rich vein of La Vizcaina in Real del Monte, and with it, the ownership of the water extracted from it. The forced labor was needed to build a tunnel known as El Aviadero, which drained the district's mines and led the water to the town of Omitlán. The waters from the mines combined to the Omitlán River and then flowed to the east, where the company built four treatment plants: the Haciendas de Sánchez, Velasco, Peñafiel, and San Miguel Regla. The water from El Aviadero was enough to be used in the plants but also by the neighboring communities. The Omitlán County took water from the Hacienda de Sánchez for the town, public buildings, cleaning of the streets, water fountains, and cattle troughs. The current then reached the Hacienda de Peñafiel and the distributor dam of La Venta, which diverted part of the water to the Hacienda de San Miguel Regla.[24]

The war of independence in 1810 in the region was in large part a rebellion against the system of exploitation of indigenous communities. During the first insurrection, indigenous leaders from Atotonilco El Grande and Huasca were the main combatants of the independentist forces and captured Pachuca and Real del Monte. During the guerrilla warfare they provided protection for the rebels that stole the silver bullions from the loyalists.[25] After Mexican independence, the company decayed, and the refining plants shut down, but the water still flowed into the communities of the Huasca Valley. The pueblos around the river used the current for their *ejidos,* their own version of the agrarian frontier. It was an agricultural land possessed and exploited collectively, located outside towns ("ejido" comes from the Latin *exitus*).

The company nonetheless retained the rights to that water. Under the liberal legislation from the Porfirian era, in the last years of nineteenth century individuals and companies, not collectives, were considered the potential users of the water. By the end of the nineteenth century, the Compañía de Real del Monte y Pachuca sent some part of the flow to the dam of the Compañía de Transmisión Eléctrica de Potencia, a company that started to provide power to the neighboring mining companies. The rest of the water flowed to the Ixtula River, a continuation of the Omitlán River, to the county of Atotonilco El Grande. There, the indigenous pueblo of Santa María Amajac and the private Hacienda El Zoquital disputed the irrigation of water.[26]

The Compañía de Real del Monte also provided water to the communities in the Pachuca Valley. The company built a treatment plant, Loreto, where the Pachuca River started its course from the mountains to the valley. The company pumped water out of the mines in the mining district, which then combined with the volumes of the Pachuca River, but they also needed water from farther away. They reached an agreement at the end of nineteenth century to pump the liquid from the cliff of Los Leones to downtown. The company provided the funds for the construction and operation of the pump. The plant received one quarter of the volume; the governmental buildings, the hospital, and public fountains could access the water at no cost. The company sold the rest of the water to private agriculture, donating the earnings to a public hospital. In sum, all the agricultural activities to the south of Pachuca depended on the volumes of water that were, to some degree, managed by the company. Haciendas, such as Coscotitlán, and pueblos, such as Mineral de Reforma and Zumpango, used the water from the river and bought some of the rights to the city of Pachuca.[27] In 1906, the United States Company acquired the Compañía de Real del Monte y Pachuca.

The American company was about to change the region's whole hydraulic system, with the arrival of the new technology of cyanidation. The local governments and the indigenous pueblos were also about to change, making a revolution for ten years.

With the Compañía de Real del Monte, the United States Company acquired rights over a considerable amount of mineral and water resources. It also had, on the other hand, a big list of pacts with neighbors about their use. One of the lawyers of the company counted at least thirty-three agreements coming from the last 150 years of operation. A change in this water distribution could affect the agricultural and power sustainability of the district, and the new manager, Morrill B. Spaulding, was not known to be a good negotiator. In 1903, while he worked in the mines in El Salvador, he entered a discussion with a physician over a bill. He destroyed the document, fought the doctor, pulled out a gun, and shot, without injuring anyone. After resisting arrest, he was put in prison, and after long negotiations, the US Consulate was able to secure his release.[28]

The personal characteristics of Spaulding and his new power in the firm quickly affected the communities that depended on this underground water. He ordered the diversion of the flow of the water from El Aviadero to Real del Monte's creek. The dam of the Compañía de Transmisión Eléctrica soon noticed the change in the volume of water received, and it was unable to provide the power it sold to the Compañía de Real del Monte. On the other hand, the ditch conducting the waters from El Aviadero flooded the town of Omitlán. Experienced Mexican officials tried to convince Spaulding to reverse the decision, but he did not take the arguments into consideration. It was not until 1908, after Spaulding was removed as manager of the Mexican division, that the officials reinstated the old arrangement of the waters.[29] By that time, the company wanted to dispose of even more water through El Aviadero tunnel. In April 1908, the new cyanidation mills in Pachuca and Real del Monte started to operate, and the cyanide solution had to find its way out from the properties of the company. The Loreto plant disposed of the water to the Pachuca River, while the solution from the Guerrero mill poured into the mines in Real del Monte and was later pumped out from the Aviadero tunnel. The users of the river in the Huasca Valley immediately noticed the change. The cattle in Atotonilco El Grande started to get sick and die, and the contaminated water destroyed the crops. The farmers from Atotonilco asked the company to reduce the length of the ditches, in order to avoid further damages. One of the lawyers of the company, Carlos Sánchez de Mejorada, believed that the damages to the pueblos could be imputable to the corporation because the river was full of "dirty and poisonous water." The legal representative in Mexico City, Salvador Cancino, proposed to divert the water before reaching Atotonilco, into the Peñafiel dam, which also would give more power to the Compañía de Transmisión Eléctrica.[30]

The company finished the new ditches in 1911. Immediately, the excess of mud in the current blocked the turbines of the Compañía de Transmisión, and by the end of the flood season the company had to divert, again, the waters to the Omitlán River, this time farther north. After the new change, the cyanide solution and the tailings polluted the pueblo of Amajac and the Hacienda El Zoquital, the property of the Craviotos, an old and important family of businessman and politicians. Simultaneously, the company faced a collective suit from farmers in Pachuca. The group, led by Vicente Aguilar, the owner of the Hacienda de Coscotitlán, demanded 22,811.25 pesos in retributions for damages in the last three years. The private demand in Pachuca revealed the mechanism of civil lawsuits that were common in the liberal Porfirian dictatorship in Mexico. In February, the engineer Cesáreo Puente, from the Ministry of Development, arrived at Pachuca to resolve the protests and find possible solutions. The diagnosis presented that spring established the existence of toxic substances in the tailings of Omitlán and Pachuca and proposed that the waters receive treatment before being liberated into the rivers.[31]

With a possible solution coming, several businessmen in Pachuca legally disputed the waters from the Pachuca River in the following months. In December, Vicente Aguilar filled a petition for the use of 660 gallons per minute of the waters that went through his lands. The petition relied on the fact that the waters would, after the treatment of the company, be free of "the volume of cyanide they carry and that makes them improper for irrigation and agricultural uses."[32] J. R. Wilson, Ignacio Urquijo, and Dante Cusi filed another petition for a small amount of water from the river, used in a cyanidation plant for tailings in the Hacienda El Pópolo, that was up the river. The amount of water was small, three gallons per second, but the possibility of another source of pollution into the river started a legal battle between the businessmen.[33]

However, these private disputes were pointless. In January 1913, the Ministry of Development approved the plans of the companies to solve the problem: the installation of three containment dams. The process was similar to the baghouses in Utah, focused on reducing the solid particles in the water, by flowing them in a circuit with settlements and filters. The gravel and sand from the tailing settled, removing the heavy metals in dry deposits. Spray and lime covered the crushed minerals, and grass and weed on top prevented the dispersion of the tailings by the wind. The water, after the settlement, overflowed through decanting and passed several presses and filters. Clarified water was then discharged into the river, still containing invisible cyanide.[34]

The trial of Aguilar and the landowners in Pachuca was not the only reason the government officials proceeded against the corporations. By the end of 1911, the Pueblo de Amajac protested the operation of the Compañía de Real del Monte to the newly appointed Ministry of Development.

> Desolation and misery will surely and relentlessly destroy our homes, because of the river water, for centuries a spring of life and prosperity for the vicinity, is now the vehicle of incalculable woes and damages in a zone over 60 miles long. . . . If the company continues to dispose the waste from the Guerrero mill to the River that soaks our land, these, as we have said, will become sterile in little time. Moreover, by inescapable consequence, a multitude of men will remain unemployed, being a real danger for the public peace, fact of real importance in the current historical times. . . . It is time to claim: the silence translates, often, into conformity, and the quietness of those who suffer encourages the powerful. The current government is an incarnation of the people, the plutocracy is dead, and the law prevails. The companies are big but just as big is the suffering people.[35]

The threat of an agrarian rebellion was not new in the history of the region, but it was especially dangerous in those last years. In the spring of 1911, the longtime dictator Porfirio Díaz, a symbol of the liberal elite integrated with international business, fled to Paris after the explosion of peasant uprisings all over the country. Revolution continued in the next ten years, with a colorful parade of old army officials, prosperous ranchers, traditional indigenous leaders, and romantic bandits taking power in Mexico City. American companies in that period were amazingly resilient in their various locations and usually could negotiate with either of the sides on the civil war, but they could not escape the revolution. The community around them, nevertheless, could not escape the corporation either.

WINCHESTER FARM, MIDVALE

In 1890 Wilford Woodruff, the president of the LDS Church, issued a manifesto that advised Mormons against plural marriage. The manifesto was a response to the disincorporation of church properties and the imprisonment of several polygamists by the federal authorities. It marked a turning point in the church's policy and allowed Utah to achieve statehood six years later. It was not, however, the first step taken by the Mormon elites to soften their relations with the federal government. One of their first strategies was the participation in the agricultural programs established by the federal state. In 1887 the Hatch Act promoted the establishment of experimental agricultural stations, and one year later, the church intervened in the installation of the Utah Agricultural Experimental Station and the agricultural college. The LDS Church gave resources to prominent Mormons to study in the East, in order to come back and be the leaders of an ambitious modernization program. They bought a house in Cambridge, where students, coming primarily from the Brigham Young Academy, lived during their grad courses at Harvard. In 1892, the University of Deseret changed its name to University of Utah, and in 1894 James Talmage, a young Mormon who did graduate studies at the University of Lehigh and Johns Hopkins, was elected president.[36] By the last years of the nineteenth century, highly educated Mormons occupied leadership positions in the educational institutions in the state and were of growing importance among the leadership of the church. When Joseph F. Smith, nephew of Joseph Smith, was elected president of the church, the agricultural ideal of the gospel had changed. While agriculture remained the basis of the Mormon family—their only way to obtain independence—knowledge replaced hard work in the accomplishment of the task on the ground.

Education was, he preached, the road "to redeem the earth and make it the Garden of Eden."[37]

This agricultural reform changed the boundary between the agricultural and the mining business in the region. The church invested in some mining companies in the south, and church officials started to work for the mining companies in Bingham. James Talmage was a geological consultant for several of the companies operating in the Wasatch Mountains. Between April and September 1905, Talmage worked for the United States Company, making several inspections of the Midvale smelter. He climbed in two occasions into the stack, collecting samples for over six hours. During that summer, he was a witness, on the side of the corporations, in the trial of *Godfrey vs.* ASARCO *et al.* At the same time, he was participating in the congressional hearings on the election of the church elder Reed Smoot to the US Senate, clarifying the current polygamist practices in Utah. Six years later he was appointed apostle of the LDS Church.[38]

This new gospel integrated the ideals of private property and industry. In the Quorum of the Twelve Apostles, Talmage was one of the first proselytizers of the Americanism of the Mormon church. He defined the LDS gospel as *Progressive Theology,* based on modern revelations that adapted continually to the spirit of times. Transformation and novelty, adaptation and rationality were the main features of the Mormonism of Talmage.[39] He was not the only one of the reformers of the church that both promoted a modern vision of Mormonism and tried to reform the agricultural culture in the area, in order to adapt to the expansion of corporate powers in the valley. In 1903, John Widtsoe and Lewis Alfred Merrill, director and agronomist of the Agricultural Experimental Station, published a report on the effects of the smoke from the smelters in the crops of the area. While the report made an exhaustive evaluation from April to October and found perceptible damages, it curtailed the demands of the farmers. The document underlined the heterogeneity of the nuisance and the temporal nature of the damage. For the most part, the report advised the farmers on how to deal with pollution, avoiding irrigation in windy days, feeding the cattle in the barn, and substituting the orchards for alfalfa. Moreover, Widtsoe and Merrill advised the farmers, "Don't ascribe all your misfortunes to the smelter smoke."[40]

As was Talmage, Merrill and Widtsoe were important reformers inside the church. Widtsoe had studied at the Brigham Young Academy, Harvard, and the University of Göttingen, all of it financed by the LDS Church. In 1900 he had become the director of the agricultural station, and, with Lewis and Merrill, he had founded the *Deseret Farmer,* a magazine committed to the reform of Mormon agriculture. From then on, he was a director, successively, of the agricultural college and the University of Utah. In 1921 he was appointed apostle and was occupied full time in theological matters. Continuing the work of Talmage, Widtsoe considered that Mormonism was a *rational theology:* it was, in principle, a material religion. There was no difference between theological and scientific reason, as the doctrinal principles were in harmony with knowledge and reason. The principal example was, for him, the arrival of the LDS Church to Utah as a revelation. The territory was, at the time, considered a desert, and divine revelation had uncovered fertility and productiveness as no other place on earth. The fact that scientists discovered the reasons for the richness of the ground was proof of the theological character of scientific knowledge. That was the testimony of the soil of God's intervention.[41]

The discovery of the properties of the ground was, in that sense, a material but also a spiritual endeavor. Lewis Merrill, even more than Talmage and Widtsoe, searched for the proof of the hand of God hidden in the dust. One of the forty-three sons of the apostle Marriner Wood Merrill, he received his education as an agronomist at the University of Utah, was a regular writer in the *Deseret Farmer,* and a big participant in the educational activities of the LDS Church. In 1903, he established six experimental dry farms in the valley with Widtsoe, a project that both of them continued for the next ten years. In 1905, when *Godfrey vs.* ASARCO *et al.* started, Merrill suggested a different strategy for the United States Company to face the suits. He considered that farmers had no training in the "modern or scientific methods in agriculture, and their judgment concerning real damage [was] not reliable."[42]

In order to show their incompetence, Merrill suggested buying the Winchester farm. The price of renting or buying, he calculated, was lower than the annual suits carried by James Winchester, greatgrandson of Benjamin. In the farm, Merrill planned to grow crops and cattle with modern methods. The United States Company hired him as the first head of the agricultural department, rented the land, and started the experiments.

Indeed, the experimental farm of Merrill was an effort to manage the limits between visible and invisible in the zone, a liminal space between corporate mining and traditional farming. The years after the trial of 1905, the corporations, the local media, and even the *Deseret Farmer,* edited by Widtsoe, changed the symbolic image of the Mormon settlers. They portrayed them as *smoke farmers,* idle workers of their land that substituted the individual work for collective suits against the companies. Incompetence and indolence were the main features of this new old Mormon settler. In comparison, the experimental farm of the United States Company publicized the hidden richness of the ground, even in the most adverse environment of the Winchester farm, located right next to the Midvale Plant. The technicians of the agricultural department took at least seventy photos per month of the farm, and their products participated in the agricultural fairs. In 1915, the United States Company won gold medals for their potatoes and apples in the Pacific International Exposition in San Diego, California. This apparent success hid a huge amount of wasted resources. The farm had nine workers and operated with a deficit of 17 percent annually.[43]

As with the Sprague method, the idea of the farm spread to other smelters in the area. In 1910 Merrill installed an experimental farm in Murray, for the ASARCO. The Guggenheim holding soon picked up the idea and installed it in several parts of the company where it faced similar conflicts. In 1914 the United States Company finally bought the Winchester farm and extended fumigation experiments over several crops: pines, wheat, barley, turnips, onions, peas, beans, tomatoes, zucchini, and Sudan grass. Merrill connected the operation of the state Agricultural Experimental Station, the six dry experimental farms in the state, the ASARCO farm, and the Winchester farm. His place in the church, moreover, helped the company negotiate with the local parish in case of grievances.[44] On his side, Sprague increased the activities in the Midvale plant. Between 1910 and 1916, the firm had increased its smelting capacity. The engineers calculated an increase in the sulfur contained in the ores, between 1914 and 1916, of 75 percent. In 1915 the United States Company installed a second baghouse to capture the rising smoke of the sulfides. Like the disposal of copper sulfides in Bingham Creek, Sprague believed that the company's growth was not detrimental to the region's agricultural development: it was its solution. He asserted that the corrosion of the fences in the regions was not the responsibility of acid rain, but the poor quality of the iron. In consequence, he promoted the use, by the vicinity, of the white lead sold by them as paint. Likewise, he blamed the grasshoppers for a big part of the damages attributed by the farmers to smelter smoke. The agricultural department started to commercialize sodium arsenide, a by-product of the operation of the baghouse, as a pesticide in the valley. By 1915, the company produced around one hundred tons of refined arsenide every month.[45]

None of this could hide the facts, and farmers continued to complain about the deterioration of their lands and crops. Many of the farmers sold or rented their properties, and the value of the civil suits dropped. With this new agricultural authority, the United States Company stopped paying reparations and won every case in court after 1908. Nonetheless, in 1914 the farmers in the vicinity started to organize a new collective suit. Under the auspice of Orrin P. Miller, a member of the presiding bishopric of the LDS Church, the farmers hired George J. Peirce, a professor of botany at Stanford University.[46]

Peirce had experience participating in big cases against metallic giants. Following decades of legal conflicts between the farmers and the smelter industry in Montana, Theodore Roosevelt's administration prepared legal cases against the Anaconda Copper Company and other firms regarding smoke damages around national forest resources. The legal team, directed by the Department of Justice, hired three professors at Stanford: Robert Swain, a chemistry professor that had testified on a previous action on behalf of the farmers; George J. Peirce; and J. P. Mitchell, another professor of chemistry. The Stanford team was in charge of collecting scientific evidence for the case on behalf of the government. The case revolved around the effects of sulfur dioxide, a by-product of the processing of sulfides, and ended with an agreement between the signature of an agreement between the government and the smelter companies in the area.[47]

Peirce examined the district in 1914 and 1915 and elaborated a report. At the same time that Alice Hamilton described sulfur dioxide as "the most painful of industrial gases" and Strong used it as a disinfectant against the spread of typhus, Peirce argued that the firms in those years only managed to hide the smoke and the more visible damages. The traditional bleaching caused by sulfur dioxide, known as burns, was reduced, but the laboratory tests and a vigilant nose could unravel the "smoke paths" on the earth. One thing that clarification by the baghouse could not eliminate was the strong sulfurous smell of sulfur oxides, and the irritation of nose, eyes, and headaches experienced by the farmers in the area. Furthermore, Peirce claimed that longtime exposure to sulfur oxides curtailed the plants' growth potential, restricting the crops. Farmers had named that phenomenon "poisoning of the soil," and the companies had denied it for ten years. By the end of 1915, a group of sixty-one farmers filed a suit against the United States Company, ASARCO, and the International Smelting Company, a subsidiary of Anaconda Copper Company that had installed a smelter in the neighboring county of Tooele. They demanded the reestablishment of the injunction of *Godfrey vs.* ASARCO *et al.*[48]

The firms were less numerous this time, but they were more powerful and better prepared for the trial. In the years since its foundation, the agricultural department of the United States Company conducted around 190 inspections every month in eighty-five different properties in the valley, collecting evidence of the negligence of the farmers. The company hired at least twenty different technicians, agronomists, entomologists, veterinarians, lawyers, experienced farmers, and stockmen, who inspected the farms of the valley and made comparisons with the results in other places in the state. All the witnesses of the US Company testified that a lack of irrigation produced the bleaching of the crops and that common diseases were responsible for the stock deaths. The trial tried to establish the existence of an invisible conquest of the company in the valley. Every technician hired by the corporation stated that they could not perceive the smoke coming out from the stacks and that only occasionally they sensed the smell, without any irritation of the eyes or nose. None of them had a headache any of those days, they declared.[49]

The company's strategy was to extend the ongoing practice, if possible, as the war boomed the prices of copper and lead and the smelters worked in a peak of production. In 1921 another technical report was obtained by the court, this time by Robert Swain, Peirce's colleague who had worked with him in the trials of Montana in 1910. Swain's arrival to the valley could not be timelier for the corporations. As the war ended, the prices of metals plummeted, and the years of 1920 and 1921 were the lowest points of production in the twentieth century in the area. There was practically no smelting; there was practically no smoke, and that impacted the results. The Swain report challenged the "invisible damage" theory of Peirce and stated that he could not find enough sulfur oxide concentration in the vicinity to be continually perceptible, except sporadically in the area of Midvale. He concluded that the Murray plant did not affect the lands of the valley, although he also suggested that further capture of pollutants should be made by the firm. In particular, he argued that the roasting and sintering plants of the US Company should be connected to the baghouse.[50] During the next decades, the corporation maintained its strategy of publicity of the Winchester farm and the disguise of the pollution in the air. Walter L. Latshaw, head of the agricultural department of the United States Company in the 1930s, described the real problem faced by the smelter and the agricultural department as the management of the limits between the visible and invisible:

> "The change we made late last summer when we initiated the practice of conducting all of our waste gases on the large brick stack, coupled with the recent change in our methods of slag disposal doing away with the accompanying smoke cloud, should do much to improve conditions in the vicinity of our smelter. Certainly, it will do much to make our neighbors less 'Midvale Smelter conscious.'"[51]

SANTA MARÍA AMAJAC, HUASCA VALLEY

In 1912, the pueblos of San Nicolás Ayotengo and Santa María Amajac, to the north of Real del Monte, joined the Sociedad Unificadora de Los Pueblos de la Raza Indígena (Unifying Society of the Peoples of the Indigenous Race), composed of at least forty pueblos in seven states. The organization had as an objective the support of the members in their demands of land restitution.[52]

The liberal governments had pushed the limits of the agrarian capitalism into the ejido lands of the pueblos, by declaring them idle and privatizing them to private haciendas in the last forty years of the nineteenth century. During the revolution, land restitution from these collective subjects was one of the central demands from all sides of the political spectrum. After three governments overthrown in four years, two groups disputed the power in Mexico: the Constitutionalists, led by the old governor Venustiano Carranza, and the Conventionalists, with the strong leadership of Pancho Villa and Emiliano Zapata. *Zapatistas* were the prototypical indigenous revolutionaries, who usually distributed land to communities after taking control over territory. Carranza combated this agrarian revolution with agrarian reform, and he signed a new agrarian law. The law recognized the rights of pueblos on their ejidos, restoring lands lost to the haciendas. That year, Zapatista militias took and redistributed land in seven different pueblos in Hidalgo. At the same time, *carrancista* forces established a local agrarian commission in Atotonilco El Grande and authorized the restitution of land to the Pueblo Santa María Amajac on the properties of the Hacienda El Zoquital, property of the Cravioto family.[53]

The ejido of Amajac now had access to land, but no water. In 1915, the agrarian commission acknowledged their demands of water restitution but offered no source of it: there appeared to be no place to find it. In 1916, the *ejidatarios* of Amajac filled new complaints against the subsidiary of the United States Company, for the pollution of the Omitlán River and the creeks around it. By then, all the fishes in the river were dead; inhabitants could not drink water, wash their clothes in the river, or bathe in the thermal waters of the community, Los Baños.[54]

The new force of the indigenous pueblos was of little concern for the corporation. In 1912, three mining companies, the Compañía de Real del Monte, the Santa Gertrudis Company, and the San Rafael y Anexas Company formed the Asociación de Compañías Beneficiadoras, Sociedad Cooperativa de Responsabilidad Limitada (Refining Companies Association, Cooperative Society of Limited Liability). Through the association, the companies faced the legal suits and jointly processed the tailings discharged into the Río Pachuca.[55] Soon, the Compañía de Real del Monte took over the full operation of the dam in Santa Julia, as it was the leading investor in the zone. Additionally, the subsidiary of the United States Company operated the other two dams, in Mineral del Chico and Omitlán, both in the hydraulic system of the Omitlán River. The company was not able in either of them to capture the cyanide in solution with the water. In the 1920s, the vice director of the subsidiary saw the control of the cyanide in the water not only impossible but also undesirable, as he claimed that the river was functioning as an open sewer for the city of Pachuca.

> The first point to take into consideration is that the cyanide in these tailings helps as an effective means of destroying the organic matter found in the River. It would be very dangerous for the health of the inhabitants of the vicinity to stop the discharge of the tailing in the River without making the arrangements to build a sewer outside of it.[56]

Soon after the beginning of the plans for the containment dams in the Pachuca and the Omitlán Rivers, the Compañía de Real del Monte y Pachuca filled the petition to obtain the concession over the metals contained in the waters. While the company entirely operated the dams in Omitlán and Atotonilco El Chico, it had to buy the rights to the tailings coming from the rest of the Society of Refining Companies in Pachuca, with the condition of assuming from them all the costs of treatment of the polluted water. Soon after the completion of the filtering plants, the company installed a zinc filter that allowed them to collect some of the solution of cyanide with gold and silver.[57] Additionally, in 1913 the subsidiary of the United States Company filed a petition to the Ministry of Development to obtain the concession to use the water treated in the Santa Julia confinement on the Pachuca River. The firm requested the water in order to conduct experiments and determine the possibilities of using cyanided water for irrigation, in several phases of treatment. In case of not using the totality of waters, they asked for the right to sell it to other users and overflow the remnants. The next years the company used the water for experimentation in irrigation in all kinds of plants, discharging the rest as they could not find any buyers. The results after one decade of tests were very disappointing. No crop could live irrigated by cyanided water, except fava bean, which grew poorly.[58]

Internalization of the processes, nonetheless, continued to be the project of the corporation, even without further state intervention. When the federal inspector visited the area of the Omitlán River, he found no sign of concern in the operation of the company. The firm had installed a three-mile pipeline from the Omitlán dam to a decanting plant in the Hacienda de Velasco. The new facility was an extended version of the dam. A closed circuit allowed them to extract more of the suspended tailings in the water, the evaporation of the cyanide solution, and the recollection of the heavy metals in ore sands.[59] In any case, the company still had to overflow around 50 percent of the cyanide solution into the Omitlán River, and the ejido of Amajac continued to demand that they divert the waters to the dam of the Compañía de Transmisión. Company officials argued that the tailings that the peasants perceived were remnants of the early operation of the company, and the analysis of the Ministry of Development found not enough contaminants in the water to harm the agricultural activities. The ejidos of the Huasca Valley complained every year for one decade of the pernicious character of the treated waters.[60]

The tailings of the containment dams posed a growing problem in the Pachuca Valley as well. The Santa Julia plant processed 2,500 tons every day, making it impossible to install a decantation circuit as in Velasco. By 1922, 8 million tons of tailings property of the US Company occupied over 250 acres south of Pachuca, which usually transformed into big clouds of sand by the strong winds of the valley. The neighboring communities around the Pachuca River formed ejidos on those years, but they never could organize a resistance as significant as the pueblos in Atotonilco El Grande. More concerned with the waste of valuable cyanide than farmers' complaints, in 1916 the company started experiments for the installation of a new process of cyanide regeneration. In the new method, the toxic water was treated with sulfide dioxide and lime, transforming the cyanide solution into hydrocyanic acid. The solution was then evaporated and condensed into sodium cyanide, and the use of filter zincs allowed obtaining gold and silver sulfides. The new process now liberated some sulfur dioxide but allowed the company to drastically reduce costs on cyanide. The firm installed a pilot plant in the Guerrero mill in Real del Monte in 1924 and finished a bigger plant in the Pachuca plant in 1930.[61] Besides the installation of the cyanide regeneration plant in Guerrero in 1924, the company had begun a process of underground centralization of the mines in the district of Pachuca and Real del Monte. Communication below the surface allowed the firm to send a more substantial part of the minerals from Real del Monte to the San Juan Pachuca shaft. From there, the Loreto plant could treat the minerals. As a consequence, the company gradually abandoned the Guerrero mill in Real del Monte until its final shutdown in 1930. In other words, less waste and cleaner water poured again from El Aviadero tunnel into the Omitlán River, by then known as *agua puerca* ("filthy water").

The ejidos in the vicinity of Omitlán immediately identified the change. This time, the ejidos of Atotonilco El Grande started to demand the rights over all the water in the river against the two companies, the Compañía de Real del Monte and the Compañía de Transmisión Eléctrica. They contested the corporate control over the body of water, as they had fought for a decade to become recognized political actors. In the summer of 1925, after the water commission declared the Agua Puerca River part of the System of National Waters, the ejidatarios of Atotonilco El Grande destroyed the pipeline from Omitlán dam to the Hacienda de Velasco plant. The destruction of that containment diverted the waters from El Aviadero back to the river and a more significant flow into the Amajac creeks. The company filed a complaint, and months later they were able to regain control over the waters, as the water commission had declared the intake illegal. The risk of the agrarian revolution was far behind, and the technical officers considered that they could use the water to experience a different version of agrarian ideals. The subsidiary of the United States Company built houses in the Hacienda de Velasco for their managers and sold the Hacienda de Sánchez to a landowner for agricultural purposes. Every year, the new properties received cleaner water, and the American engineers who lived in Velasco started to participate in the gardening and farming contests of the other officials in the subsidiary of the United States Company. The ejidos of Amajac continued to demand access to the waters from El Aviadero tunnel and the San Miguel Regla dam, still under control of the Compañía de Transmisión Eléctrica. They never succeeded.[62]

PARICUTÍN, MICHOACAN

On February 20, 1943, a volcano erupted in the cornfield of Dionisio Pulido in the town of Paricutín, in the Mexican state of Michoacán. The Purepecha farmer was working his land when he felt warmth under his feet, the earth started trembling and cracking with loud noises, and vapor erupted in several spots. Pulido alerted the population, who soon had to abandon the town. The indigenous population contemplated the violent birth of a mountain. The crater expelled incandescent rocks, bombs (globs of magma), dust, and considerable quantities of sulfurous gases. At the end of the first day of eruption, the cone measured twenty feet: by the end of the first week the volcano was 550 feet tall and, one year later, it reached over 1,000 feet. In the following days, smaller craters started to expel most of the magma, and in January 1944 another cone exploded, transforming into a different volcano: the Sapichu (small kid).[63]

The region experienced an accelerated version of the geological processes in the Bingham and Pachuca districts. Magma pushed its way to the crust, breaking it violently and relentlessly. When there was enough wind, the columns of water, silica ashes, sulfur, and carbon dioxides could cover the sky for miles, destroying the crops, collapsing roofs, and contaminating the water. In the first year, the volcano devoured the towns of Paricutín and San Juan Parangaricutiro. On May 10, 1944, the priest led a march from the church of Parangaricutiro, absorbed by the magma, to the old Hacienda de Los Conejos, for twenty miles. The governor of the state had confiscated over 2,400 acres to the neighbor ejido of San Francisco Uruapan to erect the town Nuevo San Juan, with more than four thousand refugees.[64]

Two days after the first eruption, a plane coming from Mexico City heading to Guadalajara flew low around the new crater so that the passengers could appreciate the destruction. Alan Probert, an engineer of the United States Company division in Mexico, was traveling on that flight. He had been working in the United States Company for more than twenty years, an engineer on the Midvale plant, an official in the general headquarters, and, at that time, superintendent in the Compañía de Real del Monte y Pachuca. He was heading to the Sonora mining district to evaluate some new properties, as the mines of Hidalgo were almost completely exhausted. Fascinated by the fury of the newborn peak, he and his wife Lillie spent every vacation and weekend filming the first years of the life of the volcano and the changing landscape of the Purepecha Plateau. The exodus of the inhabitants contrasted with the steady stream of visitors to the abandoned lands. American and Mexican tourists like the Proberts invaded the devastated territory. In the first months, the outlines of the buildings of the town Paricutín were visible under the mountains of ashes and tongues of lava flow. The petrified image of the town contrasted with the emergence of the volcano, moving the earth, coloring the sky and making the ground cry. While the Proberts were filming one of the strongest eruptions of the volcano, they met two Purepechans who asked if they were filming "the fire-devil." They were looking for mushrooms in the dead branches of the pines. "Nothing else grows anymore," explained one of them. "Please excuse me, *señora, pero ésta era mi milpa* [this was my little farm]," said Dionisio Pulido, who shook hands with Alan Probert.[65]

There were no casualties due to explosions or eruptions in the Purepecha region, but the cattle and inhabitants fell ill. The volcano released heavy metals, sulfur, and fine particles of silica. Sulfuric acid formed in the clouds and precipitated as rain or was produced by the contact of the ashes with the moisture in the plants. The old springs died as a consequence of the volcanic earthquakes, and heavy metals concentrated in the stream of the Itzícuaro River, the main source of water for the neighboring communities. Finally, the volcano dust, containing fine particles of silica, caused different levels of lung disease in the inhabitants of the pueblo.[66] Probert and other company engineers saw in the Paricutín an uncontrolled version of the transformation they were doing on the earth. "Never was there such a prodigal waste of energy as Mother Nature belched into the sky," observed the American engineer when recalling his memories of the emergence of the volcano.[67] One year later, in the summer of 1944, the team at the Mexican division discovered a new mineralized body below Real del Monte; they named it Veta Paricutín.[68]

By that time, the effects of the US Company's long-lasting operation revealed themselves on the territory. During the years of operation of the firm, management tried to continue their exercise of corporate sovereignty outside their formal boundaries and hide the body of waste of the Leviathan. Nonetheless, the scars of the corporate operation erupted in unexpected places.

In June and July 1939, around two hundred thousand bees died in the Salt Lake Valley. The phenomenon happened again every year from then on until 1946, coinciding with the increase of the smelters' operations to produce lead and copper for the war effort. According to the studies of the scientists of the agricultural station, the cause was a high concentration of arsenic on the surface. They found large volumes of the toxic element in the plants, earth, and water of the valley. The furnace smoke, combined with the use of arsenic as a pesticide in alfalfa and as a poison against the grasshoppers, promoted and sold by the corporation, was the main factor responsible for the high concentrations that were killing the pollinators in the summer season.[69] The farmers had denounced the poisoning of the earth for decades, but their complaints had been dismissed as ignorant and retrograde. By the 1940s, the land west of the Jordan River was no longer the installation of God on earth, but a complement of the mining operations in the area. The big mining corporation had pushed the farmers to the periphery, destroying their material base, replacing their industriousness for corporate management, and offering industrial mining as the sole mean of subsistence. Farmers in West Jordan plummeted from 30 percent of the total population in 1880 to 4.2 percent in 1940. In contrast, the workers at the mines and smelters grew from 18 percent to 70 percent in the same period.[70]

A similar transformation was suffered by the neighbors of Hidalgo. In 1895 the farmers in the Pachuca district were around 35 percent of the working adults, while mining employed 18 percent. By 1940, mining workers in Pachuca were 29 percent, while only 8 percent were farmers or peasants. In Atotonilco El Grande in 1895, agriculture employed 61 percent of the population, while in 1940 only 39 percent were working in the ejido. As there were no mines in Atotonilco, many of the peasants around the River Omitlán or Agua Puerca River migrated to work for the United States Company. People from the towns surrounding the rivers Omitlán and Pachuca occupied 15 percent of the United States Company workforce, 30 percent of all the migrants employed as laborers in the corporation.[71] These indigenous workers from the surroundings practically repeated the circuits and proportions of the repartimiento Otomís in the uprising against Romero de Terreros, two centuries before. They were still fighting for the survival of their pueblos, but their voice vanished over time. In 1946, the federal government declared valid the demands of restitution and provision of irrigation waters for the ejido of Amajac. Nonetheless, the water commission denied, once again, their access to the actual use of the stream of El Aviadero tunnel. The officials argued that there was no available volume for the irrigation of the ejido, reifying the water distribution from 1906.[72]

The changes in the communities—product of the firm's invisible limits—demonstrate that the body of waste of the corporate Leviathan was not only a failure of property rights. The long-lasting effect of corporate sovereignty outside the formal boundaries was the absorption of the subjects in the neighboring communities into the firm's formal power frontier. Visible limits could not exist and expand without the enlargement of the invisible boundaries, just as the body of minerals could not grow without the spread of the body of waste. Moreover, if the space of exception existed beyond the space of norms of the firm, its limits outlasted its legal property rights. They stayed there as proof of an otherwise ghostly existence, like scars in the land of a terrible eruption.

Epilogue

The corporation is dead; long live the corporation!

In September 1947, miners of the Paricutín mine complained to the officers of section 146 of the Mexican Miners Union, located in Real del Monte. They asked their representatives for the return of their contributions to the strike fund. The fund assured miners with resources in case of a strike over the new collective contract, #9, due the next year. The miners of Paricutín were convinced that the firm, the US Company, was about to leave the country. The rumor, widely circulated among the rank and file, was disputed by the Ministry of Labor on its meeting with the national officers of the union. Nonetheless, by then, the US Company had sold all of the Mexican subsidiary shares to Nacional Financiera (Nafinsa), a development bank of the Mexican state.[1]

The Foreign Operations Division was never a bad business for the holding. Between 1908 and 1938, it generated $45.7 million in dividends, equivalent to 46.4 percent of the total dividends delivered to the shareholders. The division's profit margin, 38.51 percent on average, was six times larger than the firm's general profit level. In the years before the Second World War, between 1935 and 1938, the Mexican subsidiary contributed 27.75 percent of the holding's total income. In those years, the depletion of reserves as a source of income in the Mexican subsidiary was considerably lower than in the rest of the divisions.[2]

Nevertheless, in 1945, for the first time the US exploration division in Mexico reported losses, as the prices of reserves fell after the war. Despite the subsidiary's good results, the holding officers decided that the drop in prices could not sustain international operation costs. The strategic decision that upper management took was to restrict their activities to a national scale and invest in increasing their share in the manufacture of metallic products. In January 1947, they sold their shares of the Mic-Mac mines in Canada, less than five years after acquiring the company's majority control, and they sold in September the Mexican division for $2,186,000. The high price they obtained for the Mexican division reduced their losses for operations to under $100,000, and in the following years, the firm slowly sold the rest of their properties inside Mexico, from Parral to Guerrero. On May 28, 1951, the US Exploration Company of Mexico, SA liquidated the last remnant of the international operation of the US Company, dividing the one hundred remaining shares, with a share value of $545, among the old management of the subsidiary and its heirs: Carlos Sánchez de Mejorada, Carlos Sánchez de Mejorada Jr., M. H. Kayser, Luis Creel Luján, and Ivan Kuryla.[3]

The end of the firm's multinational character was the preamble of a long deterioration of its operations in the United States. The sale price to the Mexican government of the division was relatively high, which allowed the firm to increase its investment into the gas and oil sector, acquiring more than eighty thousand shares of the Texas Pacific Land Trust and smaller firms in Texas and New Mexico. By the end of the 1950s, the firm had stopped operations in Alaska and started to dismantle the Midvale smelter. The firm sold some of their properties with porphyry copper deposits to the Kennecott Corporation, and the rest of their sulfide properties started to send their products to the Anaconda smelters.[4]

The lack of success in investments in the energy sector by the US Company ignited a campaign from the stockholders committee, led by Jack Wilder, for a corporate takeover. After a public confrontation, the holding management won by a small margin the election for the board of directors of 1962 and deepened their strategy of moving away from lead, silver, and gold. In 1963, the holding acquired a manufacturer in Michigan, the Mueller Brass Co., and in 1970 a producer of electrical equipment, the Federal Pacific Electric Company. In 1971 the mines in Bingham were liquidated entirely, and the firm focused on some of the zinc and copper properties of Bayard, New Mexico. The following year the US Company changed its name for the first time in seventy years to U.V. Industries. The existence of the corporate sovereign, under the new name, was by then almost wholly spectral. In 1978 the firm started its dissolution process, selling most of its assets to the Sharon Steel Corporation and liquidating the rest. The liquidation of the name represented the end of the firm's last body part, located in Maine.[5]

The history of the rise and life of the US Company was not exceptional. Nor was the history of the end of its corporate sovereignty. ASARCO sold their Aguascalientes smelter, and Phelps Dodge closed its Nacozari plant in Sonora, almost two decades before the US Company sold its properties to the Mexican government. After the war, facing the low tendency in metals' prices, mining corporations diversified and integrated their operations. By the 1960s, Anaconda, Phelps Dodge, and Kennecott had divisions manufacturing copper and lead products, especially electrical equipment. The new divisions changed the nature of their exploitation, and some of the companies had organizational reforms: while the US Company transformed into U.V. Industries, Phelps Dodge changed to Phelps Dodge Industries, manufacturing metallic equipment on the West Coast. Kennecott also invested in Texas oil, while Phelps Dodge invested in uranium in the Southwest. Nevertheless, they never transformed into large actors again. In 1965 a Mexican firm, Grupo Mexico, acquired 49 percent of ASARCO's shares, effectively controlling the executive committee. In 1971, Anaconda Copper sold the biggest copper mine in Mexico, Cananea, to the Mexican government. Simultaneously, the Chilean congress approved the copper sector's nationalization affecting their properties in Chuquicamata. Two years later, Anaconda Copper supported the coup against Allende, but they never received the division back, and in 1977 the firm was absorbed by Atlantic Richfield. Finally, after almost ten years of total stoppage of operations, in 1987, British Petroleum took control of the rest of Kennecott's holdings.[6]

During the decades of operating as corporate Leviathans, the mining giants had implemented a global sovereignty system that conveyed the earth's depths across the continent to the metallic markets in New York. The extraction of "raw" materials entailed technological and material flows in a complex global value chain. The growing demand for new ores entailed a territorial growth of the companies from east to west, to the north and south. The most mineralized areas soon became hubs of multinationals with industrial mines and modern smelters. As the companies consolidated their operations, divisions became more dependent on each other to be profitable and transformed into departments. Railroads, selling departments, supplies divisions, and research laboratories operated as successive layers in the mining multinationals' divisions and parent company. Minerals, technology, and technicians circulated across the corporations' map, established their own rules, and exercised sovereign power over the territories.

This book has traced the development of the United States Company as one of these sovereign titans. While the idea of the Leviathan entails the coordination of thousands of bodies into a single organization, the history of the US Company reflected the multiplicity at the inside and the changing nature of the organization.

The material reality of the corporation required the location of mineral bodies in the surface and the underground, in the metallic mountains, as well as their extraction, processing, and refining into valuable metals. It also embedded the construction of a body of waste, the most voluminous component of the firm: an ever-growing invisible limit that mirrored the creation of wealth by producing valuable compounds. The location of deposits connected the firm's history with the past, the eons that produced the hard rock mountains. However, the disposal of waste connected the corporation with the future. The ruins of the firm's operation stayed in the physical bodies of the inhabitants of the mining communities.

The corporation's organization, its Form, was not empty and abstract, but was filled with concrete historical actors: the technical officers. The firm's incorporation and the formal internal rules created in 1906 delineated an abstract organization with a centralized volition but a ghostly material existence. Over time, the firm's officials became more and more aware of their role as magistrates in a non-territorial kingdom. They shared collective corporate consciousness, shared and administrated authority in the firm, and homologated their bodily experiences in the divisions. The ghosts of their participation in this corporate sovereign followed them for years, and in the coming decades they made sure to write their history in the firm and their connections with the past of other corporate bodies.

The workers in the firm experienced transformation in those decades. They were, in the beginning, subjects of the firm on an individual level, through independent contracts, and their bodily experiences were isolated in the corporation's eyes. Throughout the years, they built a collective on their own, a body of subjects, that organized their struggles. Their new organizations also focused on their bodily experiences, and in finding this collective, the body of the subjects of the mining company also changed. Male bodies were the main protagonists of the workshops, mountains, and plants, but female bodies became the main drivers of community life and union mobilization in the middle of the twentieth century.

As with any other sovereignty, the proof of the firm's unity was its change over time and, eventually, its collapse. The other sovereign power, the nation-state, recognized corporate extraterritorial sovereignty, and the corporate, national, and imperial sovereignty structures modified themselves over those years. The shifting economy of the continent, from a highly integrated regional value chain to a nationally centered market, was lethal for most of them. The increase in scope and further integration of activities, now on a national level after the Second World War, did not modernize or improve their organizations but destroyed them. They never recovered their profitability even when manufacturing consumer goods, and they could not disrupt the market of established actors in the energy business. Their power vanished, little by little, during the Cold War era, but their demise was not the end of corporate sovereigns. They were quickly replaced by new manufacturing firms and, later, by communications and social network behemoths that expanded their rule over the continent.

The year after the sale of the Compañía de Real del Monte y Pachuca to Nafinsa, the US Exploration Company sold some of the adjacent properties in the area. In 1948, the new administration sold the old Hacienda de San Miguel Regla to the Pan-American Doctor's Club. The old Hacienda de Beneficio, flooded by the dam of San Antonio Regla, was transformed into a hotel by the new owners. American tourists could now swim in isolated pools, use the basaltic prisms' thermal waters, and enjoy the mining industry's ruins. The old furnaces' stacks emerged in the middle of a lake, repopulated by some fishes, birds, and trees. In the following decade, the Mexican government gave the new American owners subsidies and technical assistance to install a small airport and promote agriculture in the zone to meet the resort's needs. The propaganda for this postindustrial paradise explained the extraordinary and peaceful decay: "For a while, the Huasca Valley was a thriving, well-irrigated agricultural area, but the beginning of the revolution in 1914 caused its abandonment."[7]

The sale revealed two tendencies in the fate of multinationals after the mining corporations' half-century reign in the area. In the first place, the sale of the old amalgamation hacienda in Huasca reflected the persistence of the space of exception of the firm, the body of waste, over the other bodies of the corporate sovereign. Even after the end of the property rights of the US Company, the ejido could not obtain access to the waters and stayed dependent on the excess of water of the Pan-American club. This water still contained the imprint of the corporate body of waste, and the dust of the Leviathan's ruins still flew into the lungs of the inhabitants. The Huasca Valley neighbors continued, over the following years, to have critical health issues, especially tuberculosis.[8]

The city slowly absorbed the other end of the company's waste, the Santa Julia tailings, and the dusty ground became part of the population's living environment. The old tailings are now the ground of the neighborhoods of Santa Julia and Ampliacion Santa Julia, as well as some upper-middle-class suburb developments. Children play soccer, a heritage from the English migration in the area, in contained tailings that still cover the players in fine dark dust. From time to time, the residents of Pachuca can see bulldozers reclaiming the dust of some vacant plot, shipping it for a new round of refinement of the traces of silver and gold. In 1993 Adolfo Lugo Verduzco, governor of Hidalgo and nephew of the former governor and mining worker Bartolomé Vargas Lugo, inaugurated the Estadio Hidalgo for the Pachuca Soccer Club. The building was erected on old mining tailings, and the local press underlined that the traces of silver and gold in the ground made it the most valuable stadium in the country. They neglected to mention the traces of mercury, cyanide, and heavy metals that also fly on many days on Pachuca, *la bella airosa* (the windy beauty).[9]

The space of exception of the United States Company extended temporally, this time, after the end of the mining operations in other firm locations. In Alaska, the camps' abandonment left the artificial lakes and dredges to the corrosion of time. To this day, the Chatanika camp remains a rusty memory of the days of exploitation, with a dredge full of graffiti and the mountains of tailings naked of any kind of vegetation. The East and South Chicago areas struggle, to this day, with lead and manganese pollution. In 2016, the West Calumet Housing Complex residents discovered that their houses lay on the Anaconda smelter's old terrain and the US Lead Refinery. The announcement of an imminent relocation of the residents, primarily people of color, came five years after the place was named an EPA Superfund site with the discovery of hundreds of lead poisoning cases in children. Just in the house of one of the residents, the EPA found levels eighty times higher than the federal safety limit. Between 2005 and 2015, the Indiana State Department of Health found at least 160 children under five with lead levels above the federal health guidelines; almost three out of every ten kids surveyed.[10]

In the second place, the sale of the old hacienda in San Miguel Regla showed firms' capacity to occupy the space of old sovereigns. American foreign investment in Latin America experienced a second wave after the Second World War, this time with manufacturing corporations opening factories in the region. New corporate Leviathans were able to adapt to the new sovereignty of national states south of the Río Bravo, and the growth of tourism from the United States increased the flow of Americans into Mexico. Sulfides and mercury stayed in the old hacienda, but the hotel continues, to this day, to be one of the main attractions for foreigners in the old mining district.

In other words, if the body of waste of the United States Company remained in several forms in the territory, corporate bodies found ways to establish themselves on top of the old Leviathans' ruins. After its abandonment in 1971, the Midvale site remained an inhospitable part of the Salt Lake Valley. The Environmental Protection Agency started cleanup actions in the site in 1990, inaugurating a three-decade rehabilitation of the land and river. In 2018, after millions of dollars of investment by the federal agency, the Zion Bank announced the construction of a mixed development in the area. The new corporate project consists of a tech campus for more than 2,000 employees of the firm, retail outlets, and 3,500 residential units on the 265 acres of the smelter's old terrain.[11] The old mining Leviathan was dead, but from its ruins resurrected a new corporate sovereign.

Notes

INTRODUCTION

1. "Copper," CRB *and Mining Outlook*, February 7, 1917.

2. USSRM Co, *Annual Report*, 1932–38.

3. Alfred D. Chandler Jr., *The Visible Hand* (Cambridge, MA: Harvard University Press, 1993), appendix A, 503–12.

4. David Kirkpatrick, *The Facebook Effect: The Inside Story of the Company That Is Connecting the World* (New York: Simon and Schuster, 2011), 254.

5. Blue Origin, "Going to Space to Benefit Earth," full event replay available at youtu.be/GQ98hGUe6FM.

6. Edmund Burke, *The Works of the Right Honourable Edmund Burke*, 7th ed., vol. 9 (Boston: Wells and Lilly, 1826), 354.

7. Burke, *Works of the Right Honourable Edmund Burke*, vol. 9, 349.

8. Marcelo Bucheli, "Multinational Corporations, Totalitarian Regimes and Economic Nationalism: United Fruit Company in Central America, 1899–1975," *Business History* 50, no. 4 (July 2008): 433–54, https://doi.org/10.1080/00076790802106315.

9. CIA Archives, CIA-RDP74B00415R000300020011–1.

10. Thomas F. O'Brien, *The Revolutionary Mission: American Enterprise in Latin America, 1900–1945* (Cambridge: Cambridge University Press, 1999).

11. "The king was no more a corporator of Rhode Island than he was a corporator of the city of Norwich or of the East India Company." Frederic W. Maitland, "Crown as Corporation," *Law Quarterly Review* 17 (1901): 131.

12. Joshua Barkan's *Corporate Sovereignty: Law and Government under Capitalism,* published by the University of Minnesota Press in 2013, and Grietje Baar's *The Corporation, Law and Capitalism,* published by Brill in 2019, explore the legal characteristics of the modern corporation and sketches an intellectual history of personhood, property, criminality and citizenship. *The Corporation: A Critical, Multi-Disciplinary Handbook,* edited by Grietje Baars and Andre Spicer, published by Cambridge University Press in 2017, offers forty small introductions into several general and specialized debates over the form of the corporation in half a dozen of different disciplines.

13. Huw Bowen's *The Business of Empire: The East India Company and Imperial Britain, 1756–1833* (Cambridge: Cambridge University Press, 2005) and Philip J. Stern's *The Company-State: Corporate Sovereignty and the Early Modern Foundations of the British Empire in India* (Oxford: Oxford University Press, 2011) describe the example of the East India Company charter as an example of sovereign firms during the formation

of nation-states and empires. The forthcoming book of Elizabeth Cross, Georgetown University, analyzes the French East India Company in similar optics, emphasizing the long-term effects of the merchant politics of the firm.

14. The case study on the private sovereignties in Africa in the nineteenth century follows the same pattern of using the legal basis as analytical framework. Steven Press, *Rogue Empires* (Cambridge, MA: Harvard University Press, 2017).

15. The studies on the United Fruit Company follow the operations of the firms in different nations. For Honduras, John Soluri, *Banana Cultures: Agriculture, Consumption, and Environmental Change in Honduras and the United States* (Austin: University of Texas Press, 2005); for Colombia, Marcelo Bucheli, *Bananas and Business* (New York University Press, 2005); for Costa Rica, Aviva Chomsky, *West Indian Workers and the United Fruit Company in Costa Rica, 1870–1940* (Baton Rouge: Louisiana State University Press, 1996); for Costa Rica and Guatemala, Jason M. Colby, *The Business of Empire: United Fruit, Race, and US Expansion in Central America* (Ithaca, NY: Cornell University Press, 2011); for Ecuador, Steve Striffler, *In the Shadows of State and Capital* (Durham, NC: Duke University Press, 2001); and for a discussion between most of these scholars, see Steve Striffler et al., eds., *Banana Wars: Power, Production, and History in the Americas* (Durham, NC: Duke University Press, 2003).

16. The best documented research on the U. Fruit Company, Colby's book, relies on the records of the Northern Railway Company, who contained the correspondence of the division in Costa Rica. Buchelli's study on the division in Colombia only uses a dozen of monthly letters, monthly reports (1948–68) recording mostly profits and losses, and the annual reports (1900–1970). Striffler's book on the firm in Ecuador had no access to any corporate archive of the division.

17. Similar issues face the general histories of extractivism in the Americas at the time. See Myrna L. Santiago, *The Ecology of Oil: Environment, Labor, and the Mexican Revolution, 1900–1938* (Cambridge: Cambridge University Press, 2006); Thomas Miller Klubock, *Contested Communities* (Durham, NC: Duke University Press, 1998); and Angela Vergara, *Copper Workers, International Business, and Domestic Politics in Cold War Chile* (University Park: Penn State Press, 2010). Examples of the ideological nature of American firms in the world, although in a very general way, can be found in Victoria de Grazia, *Irresistible Empire: America's Advance through Twentieth-Century Europe* (Cambridge, MA: Harvard University Press, 2009); and Thomas F. O'Brien, *The Revolutionary Mission: American Enterprise in Latin America*, vol. 81 (Cambridge: Cambridge University Press, 1900).

18. Patricia Springborg, "Leviathan, the Christian Commonwealth Incorporated," *Political Studies* 24, no. 2 (1976): 171–83.

19. Thomas Hobbes, *Leviathan: Or the Matter, Form, and Power of a Commonwealth Ecclesiastical and Civil* (London: G. Routledge and Sons, 1887), 221.

20. Robert Fredona and Sophus A. Reinert, "Leviathan and Kraken: States, Corporations, and Political Economy," *History and Theory* 59, no. 2 (2020): 167–87.

21. Chandler uses the metaphor of firms to exemplify their nature as artifacts.

Alfred Chandler and Bruce Mazlish, "Introduction," in *Leviathans: Multinational Corporations and the New Global History*, edited by Alfred D. Chandler and Bruce Mazlish (Cambridge: Cambridge University Press, 2005).

22. Hobbes, *Leviathan*, 114.

23. Thomas Hobbes, *The Elements of Law, Natural and Politic: Part I, Human Nature, Part II, de Corpore Politico; with Three Lives* (New York: Oxford University Press, 1999), 93.

24. Ronald Harry Coase, "The Nature of the Firm," *Economica* 4, no. 16 (1937): 386–405.

CHAPTER 1. THE GHOST AND THE MACHINE

1. Henry David Thoreau, *The Maine Woods: A Fully Annotated Edition* (New Haven, CT: Yale University Press, 2009), 5.

2. Thomas Lynch, "The 'Domestic Air' of Wilderness: Henry Thoreau and Joe Polis in the Maine Woods," *Weber Studies* 14, no. 3 (1997): 38–48; and Micah Pawling, "Wabanaki Homeland and Mobility: Concepts of Home in Nineteenth-Century Maine," *Ethnohistory* 63, no. 4 (2016): 621–43.

3. Thoreau, *Maine Woods*, 276.

4. Michael C. Connolly, *Seated by the Sea: The Maritime History of Portland, Maine, and Its Irish Longshoremen* (Gainesville: University Press of Florida, 2010), 1–84.

5. USSRM Co., *United States Smelting, Refining, and Mining Company . . . : Preferred Stock, Common Stock* (Boston: USSRM, 1916).

6. Henry S. Williams and Herbert E. Gregory, "Contributions to the Geology of Maine," *Bulletin*, no. 165 (1900).

7. *International Mining Manual*, Western Mining Directory Company, New York, 1907, 114.

8. *Santa Clara County vs. Southern Pacific Railway Co.*, 118 US 394 (1886).

9. United States et al., *Amendment of Sherman Antitrust Law: Hearings before the United States Senate Committee on the Judiciary, Sixtieth Congress, First Session, on Apr. 23, 1908* (Washington, DC: US Government Printing Office, 1971).

10. Herbert Milton Heath, *Comparative Advantages of the Corporation Laws of All the States and Territories* (Maine: Kennebec Journal Print, 1902).

11. *Business Corporations under the Laws of Maine*, Making of Modern Law Legal Treatises (Portland, ME: Corporation Trust Company of Maine, 1903); American Mining Manual, 1907.

12. Congreso de los Estados Unidos Mexicanos, "Código de Comercio," October 7, 1889.

13. Peter F. Drucker, *Concept of the Corporation* (New Brunswick, NJ: Transaction Publishers, 1993), 38–50.

14. Gilbert Ryle, *The Concept of Mind* (Routledge, 2009), 2009.

15. *Milford News*, "Bingham's First Billion," November 2, 1937, 4; *Salt Lake Herald*, "Sixty Stamp Mill to Be Erected," March 25, 1880, 3; *Salt Lake Herald*, "New

Companies," May 1, 1880, 3; Casper A. Nelson, "Smelters," *Midvale History, 1851–1979* (Midvale, UT: Midvale Historical Society, 1979), 232.

16. *Salt Lake Herald*, "Resolutions on the Question Silver," January 24, 1886, 12; *Salt Lake Herald*, "The UP Takes a Big Hand," March 8, 1891, 8; *Salt Lake Herald*, "Interview with Prof. Holden," May 16, 1886, 9; *Salt Lake Tribune*, June 19, 1892, 9; *Salt Lake Herald*, "Save Him," March 26, 1886, 4; *Salt Lake Herald*, October 2, 1896, 8; National Mining Congress, ed., *Report of the Proceedings* (Denver); USSRM Co., *Smelting History* (USHS, MSS A-6028 c.1,); and *Salt Lake Tribune*, "University Notes," April 20, 1891, 8.

17. See Jerre C. Murphy, *The Comical History of Montana: A Serious Story for Free People: Being an Account of the Conquest of America's Treasure State by Alien Corporate Combine, the Confiscation of Its Resources, the Subjugation of Its People, and the Corruption of Free Government to the Uses of Lawless Enterprise and Organized Greed Employed in "Big Business"* (San Diego: Scofield, 1912), 2.

18. Christopher Powell Connolly, *The Devil Learns to Vote, the Story of Montana* (New York: Covici, Friede, 1938); *New York Times*, "United Metals Selling Company Will Win Control over Heinze," March 27, 1906; and Kenneth Ross Toole, "The Anaconda Copper Mining Company: A Price War and a Copper Corner," *The Pacific Northwest Quarterly* 41, no. 4 (1950): 312–29.

19. Michael P. Malone, *The Battle for Butte: Mining and Politics on the Northern Frontier, 1864–1906* (Seattle: University of Washington Press, 2012), 199.

20. Leonard J. Arrington and Gary B. Hansen, *The Richest Hole on Earth: A History of the Bingham Copper Mine*, vol. 11 (Logan: Utah State University Press, 1963), 15–28; Elizabeth W. Lind, "Mills," in *Midvale History, 1851–1979*, ed. Maurice C. Jensen (Midvale, UT: Midvale Historical Society, 1979), 217–23.

21. James E. Fell Jr., *Ores to Metals: The Rocky Mountain Smelting Industry* (Boulder: University Press of Colorado, 2009), 201–54.

22. See an account of the consolidation process in Malone, *Battle for Butte*.

23. Isaac Frederick Marcosson, *Metal Magic: The Story of the American Smelting & Refining Company* (New York: Farrar, Straus, 1949), 57–69.

24. John Mason Boutwell, *Economic Geology of the Bingham Mining District, Utah* 38 (Washington, DC: US Government Printing Office, 1905).

25. *Salt Lake Herald*, "Save Him," March 26, 1886, 4; *Salt Lake Herald*, October 2, 1896, 8; National Mining Congress, *Report of the Proceedings*; USSRM Co, *Smelting History*; *Salt Lake Tribune*, "University Notes," April 20, 1891, 8; *Salt Lake Tribune*, February 5, 1897, 8; and *Ogden Standard*, "Death Notices," August 27, 1913.

26. Arrington and Hansen, *Richest Hole on Earth*, 29–50; Marcosson, *Metal Magic*, 147.

27. USSRM Co., *Smelting History*; USSRM Co., *United States Smelting, Refining, and Mining Company . . .*; SLCR, Miscellaneous, roll 140, *West Mountain Mining District Proofs of Labor*, 1936. In 1916, the company had 134 patents in the area of Bingham, twenty-six mines in Eureka, Utah, and seventy-six in Ruby Hill, Nevada; USSRM, "Preferred and Common Stock"; *Salt Lake Tribune*, December 20, 1899.

28. WSJ, June 30, 1920, 7. *NY Standard*, "Silver Producing"; Thomas W. Dombroski,

How America Was Financed: The True Story of Northeastern Pennsylvania's Contribution to the Financial and Economic Greatness of the United States of America (Bloomington, IN: iUniverse, 2011); *The Bakers Magazine,* vol. 94, 82–84.

29. Miguel O. de Mendizábal, "Los Minerales de Pachuca y Real Del Monte En La Época Colonial: Contribución a La Historia Económica y Social de México," *El Trimestre Económico* 8, no. 30 (2 (1941): 253–309.

30. Robert W. Randall, *Real Del Monte: A British Silver Mining Venture in Mexico,* vol. 26 (Austin: University of Texas Press, 1972).

31. Rocío Ruiz de la Barrera, "La Empresa de Minas Del Real Del Monte (1849–1906)" (PhD diss., El Colegio de México, 1995), 134–53.

32. Marcosson, *Metal Magic,* 53.

33. Isaac Frederick Marcosson, *Anaconda* (New York: Dodd, Mead, 1957), 255.

34. Jesús Gómez Serrano, *Aguascalientes: Imperio de Los Guggenheim* (México: FCE, 1982), 49; Marvin D. Bernstein, *The Mexican Mining Industry, 1890–1950: A Study of the Interaction of Politics, Economics, and Technology* (New York: State University of New York, 1965), 63; and Edward Beatty, *Technology and the Search for Progress in Modern Mexico* (Oakland: University of California Press, 2015), 145–43.

35. Ruiz de la Barrera, *La Empresa de Minas Del Real Del Monte (1849–1906),* 333–36.

36. *Nevada State Journal,* "Big Mexican Mining Deal," February 17, 1906.

37. CEHM, CDLIV, 2a, 1905, box 10, Legajo 65.

38. CEHM, CDLIV, 2a, 1906, box 3, Legajo 88.

39. EMJ, June 2, 1906, 1067; *Los Angeles Express,* May 26, 1906, 5.

40. International Union of American Republics, *Monthly Bulletin of the International Bureau of the American Republics,* Washington, DC, vol. 16, April 1904, 1231–32; vol. 21, July 1905, 1348.

41. Beatty, *Technology and the Search,* 200, 280.

42. Bernstein, *Mexican Mining Industry,* 112–13; Gómez Serrano, *Aguascalientes,* 225–48; AHCRMyP, FN, JD, Actas e Informes, vol. 1, file 2; *Annual Report,* 1908–13.

43. Gómez Serrano, *Aguascalientes,* 31–32; Bernstein, *Mexican Mining Industry,* 143–48.

44. USSRMCo, *Annual Report,* 1920–28; AHCRMyP, FN, JD, Actas e Informes, vol. 1, files 1–2, p. 7; AHCRMyP, FN, JD, vol. 3, file 16; USSRMCo, *Annual Report,* 1914–17.

45. USSRM, "Preferred and Common Stock"; AHCRMyP, JD, Actas e informes, vol. 1, file 2.

46. Javier Ortega Morel, "Minería y tecnología: La compañía norteamericana de Real del Monte y Pachuca, 1906 a 1947" (PhD diss., Facultad de Filosofía y Letras, Universidad Nacional Autónoma de México, 2010), 37; AHCRMyP, FN, correspondencia, M. H. Kayser, vol. 23, file 6; AHCRMyP, FN, JD, vol. 3, file 16.

47. MLUU, Gaylon Hansen papers, boxes 1–8, passim; USSRM, Annual Report, 1906–47, passim.

48. AHCRMYP, FN, AED, Venta de Metales, vol. 89, file 72; *Mining Reporter,* October 23, 1901, 334; *Salt Lake Evening Telegram,* February 10, 1910; F. Ernest Richter, "The

Amalgamated Copper Company: A Closed Chapter in Corporation Finance," *The Quarterly Journal of Economics* 30, no. 2 (1916): 387–407.

49. Marcosson, *Metal Magic*, 53.

50. USSRM, *Ax-I-Dent-Ax*, July 1929, 5, 15; USSRM, *Ax-I-Dent-Ax*, April,1921, 17.

51. Richter, "Amalgamated Copper Company"; Federal Writers' Project Indiana, *The Calumet Region Historical Guide: Containing the Early History of the Region as Well as the Contemporary Scene within the Cities of Gary* (East Chicago: German Printing Company, 1939), 56.

52. *The Mining American*, 1912, vol. 65, 349.

53. USSRMCo, *Annual Report*, 1918–36; USSRM, *Ax-I-Dent-Ax*, September 1931, 17.

54. Anaconda Copper Mining Company, "Report of the Anaconda Copper Mining Company," *Report of the Anaconda Copper Mining Company*, 1906–14 passim; Kennecott Copper Corporation, *Annual Report of the Kennecott Copper Corporation*, 1916–33 passim.

55. As in Negri's and Hardt's terms, Michael Hardt and Antonio Negri, *Empire* (Cambridge, MA: Harvard University Press, 2000), 33; USSRM Co, *Annual Report,* 1910.

56. Arrington and Hansen, *Richest Hole on Earth*, 29–50; Marcosson, *Metal Magic*, 147.

57. "Informe anual de la Comisión de Yauli," Boletion del cuerpo de Ingenieros de Minas del Perú, 1908; *The Mining Reporter*, July 11, 1907, 22; EMJ, May 25, 1907, 993; *Annual Report*, USSRM, 1907.

58. EMJ, vol. 97, no. 10, 526.

59. *Mining Science*, vol. 62, November 10, 1910, 450; Susan Ewing, *Resurrecting the Shark: A Scientific Obsession and the Mavericks Who Solved the Mystery of a 270-Million-Year-Old Fossil* (New York: Simon and Schuster, 2017); Walter Frederick Ferrier, *Catalogue of a Stratigraphical Collection of Canadian Rocks Prepared for the World's Columbian Exposition, Chicago, 1893*, vol. 3 (Ottawa: Geological Survey of Canada, 1893); and EMJ, December 24, 1898, vol. 66, 764.

60. The Compañía Exploradora had a capital of only 100,000 pesos, and 565,600 pesos in debts with the US Smelting Exploration, SA. AHCRMyP, FN, Jurídico, Representación, asociación y constitución, Compañías, asociaciones y negociaciones, vol. 1, file 2.

61. These included the British Guano Exploration Company and the British Transvaal Exploration Syndicate.

62. USF Co., *Thirty Years of Coal* (Salt Lake City, 1946); USSRMCo, *Annual Report*, 1911.

63. *Mining and Engineering World*, vol. 32, 223; AHCRMyP, correspondencia, M. H. Kuryla, vol. 27, file 47; AHCRMyP, FN, Sección Jurídico, Serie Representación, asociación y constitución, Subserie, Compañías, asociaciones y negociaciones, vol. 9, p. 13; AHCRMyP, FN AED, Exploracion, vol. 108; AHCRMyP, Jurídico, Propiedades, contratos y denuncios; denuncios y concesiones, vol. 1, file 3.

64. International Bureau of the American Republic, *Monthly Bulletin*, 1905, 1408.

65. AHCRMyP, FN, correspondencia, A. B. Marquand, vol. 32, file 39. FN, AED, Exploración, vol. 111, exp. 190, vol. 111, exp. 191. AHCRMyP, FN, correspondencia,

M. H. Kayser, vol. 23, file 7; FN, AED, exploracion, vol. 106, exp. 90; AED, Admnistracion Oficina Boston, vol. 61, file 25.

66. USSRM, *Ax-I-Dent-Ax,* January 1930, 31; AHCRMyP, FN, AED, Exploracion, vol. 111, file 192.

67. Richard E. Lingenfelter, *The Hardrock Miners: A History of the Mining Labor Movement in the American West, 1863–1893* (Berkeley: University of California Press, 1981), 4–5.

68. Charles Caldwell Hawley, *A Kennecott Story: Three Mines, Four Men, and One Hundred Years, 1887–1997* (Salt Lake City: University of Utah Press, 2015), 49–60.

69. Arthur Coe Spencer and Charles Will Wright, *The Juneau Gold Belt, Alaska* (Washington, DC: US Government Printing Office, 1906); USSRMCo, *Annual Report,* 1911–18.

70. Alaska Mining Hall of Fame, "Wendell P. Hammon," from Alaskamininghallof fame.org/inductees/hammon.php.

71. USSRMCo, *Annual Report,* 1911–30.

72. Coase, "Nature of the Firm," 37–54.

73. Hobbes, *Elements of Law,* 81.

74. Karl Marx, *Capital,* trans. Ernest Untermann, vol. I (Chicago: Kerr, 1912), 784–838.

CHAPTER TWO. THE BODY OF THE SUBJECTS

Some paragraphs of this chapter had a previous version of García Solares, "Striking Hard Rock Veins. Multinational Corporations and Miners' Unions in Mexico and the United States, 1906–1952."

1. 67th meeting of the Boston Section of the AIME, MIT Outstanding Collections, MC 0094.

2. de Mendizábal, "Los Minerales de Pachuca."

3. Doris M. Ladd, *Génesis y desarrollo de una huelga: Las luchas de los mineros mexicanos de la plata en Real del Monte, 1766–1775* (México: Alianza Editorial, 1992), 23–39, 75–107, 163–186; Noblet Barry Danks, "The Labor Revolt of 1766 in the Mining Community of Real Del Monte," *The Americas* 44, no. 2 (1987): 143–65; and Rodrigo Perujo de la Cruz, "Al grito de ¡revoltura!: Rebelión y cultura política en Real del Monte en 1766" (bachelor's thesis, Facultad de Filosofía y Letras, Universidad Nacional Autónoma de México, 2012), 7–70.

4. Robert W. Randall, "British Company and Mexican Community: the English at Real del Monte, 1824–1849," *The Business History Review* (1985): 622–44.

5. Alan Derickson, "Health Programs of the Hardrock Miners' Unions, 1891–1925" (PhD diss., University of California, San Francisco, 1986), 18–21.

6. *The Penny Magazine,* December 27, 1834, 500.

7. Price, "'West Barbary;' or Notes on the System of Work and Wages in the Cornish Mines," *Journal of the Royal Statistical Society* 51, no. 3 (1888): 494–566; Roger Burt

and Sandra Kippen, "Rational Choice and a Lifetime in Metal Mining: Employment Decisions by Nineteenth-Century Cornish Miners," *International Review of Social History* 46, no. 1 (2001): 45–75.

8. Davies, "Cornish Miners and Class Relations in Early Colonial South Australia: The Burra Burra Strikes of 1848–49," *Australian Historical Studies* 26, no. 105 (1995): 568–95.

9. Lingenfelter, *Hardrock Miners,* 6–31.

10. A combination of piecework and value-based payment persisted in most copper, silver, and gold operations. Frederick Wolf, Bruce Finnie, and Linda Gibson, "Cornish Miners in California: 150 Years of a Unique Sociotechnical System," *Journal of Management History* 14, no. 2 (2008): 144–60; Arthur Cecil Todd, *The Cornish Miner in America: The Contribution to the Mining History of the United States by Emigrant Cornish Miners—the Men Called Cousin Jacks,* vol. 6 (Glendale, AZ: Arthur H. Clark Company, 1967).

11. Philip J. Mellinger, *Race and Labor in Western Copper: The Fight for Equality, 1896–1918* (Tucson: University of Arizona Press, 1995), 5–6, 17–90; Andrew B. Arnold, *Fueling the Gilded Age: Railroads, Miners, and Disorder in Pennsylvania Coal Country,* vol. 2 (New York: New York University Press, 2014), 63–153; and Mark Wyman, *Hard Rock Epic: Western Miners and the Industrial Revolution, 1860–1910* (Berkeley: University of California Press, 1989), 160–65.

12. Derickson, "Health Problems," 101–24; Alan Derickson, *Workers' Health, Workers' Democracy* (Ithaca, NY: Cornell University Press, 2019), 125–54.

13. Much research on transnational workers' organizations describes networks created by ethnic and national groups migrating around the globe, or militant plots with centralized strategies. See Donna Gabaccia, Franca Iacovetta, and Fraser Ottanelli, "Laboring across National Borders: Class, Gender, and Militancy in the Proletarian Mass Migrations," *International Labor and Working-Class History* 66 (2004): 57–77; Christian Koller, "Local Strikes as Transnational Events: Migration, Donations, and Organizational Cooperation in the Context of Strikes in Switzerland (1860–1914)," *Labour History Review* 74, no. 3 (2009): 305–18; Brenda Baletti, Tamara M. Johnson, and Wendy Wolford, "'Late Mobilization': Transnational Peasant Networks and Grassroots Organizing in Brazil and South Africa," *Journal of Agrarian Change* 8, nos. 2–3 (2008): 290–314; and Nicola Pizzolato, "Workers and Revolutionaries at the Twilight of Fordism: The Breakdown of Industrial Relations in the Automobile Plants of Detroit and Turin, 1967–1973," *Labor History* 45, no. 4 (2004): 419–43. These explanations often complement traditional analysis of labor transnationalism as a contentious policy with dispersed and episodic behavior. See Leopold H. Haimson and Charles Tilly, eds., *Strikes, Wars, and Revolutions in an International Perspective: Strike Waves in the Late Nineteenth and Early Twentieth Centuries* (Cambridge: Cambridge University Press, 2002); Carol Conell and Samuel Cohn, "Learning from Other People's Actions: Environmental Variation and Diffusion in French Coal Mining Strikes, 1890–1935." *American Journal of Sociology* 101, no. 2 (1995): 366–403; and Thomas C. Buchanan, "Class Sentiments:

Putting the Emotion Back in Working-Class History," *Journal of Social History* 48, no. 1 (2014): 72–87.

14. *Iron Trade Review*, vol. 58, June 1916, 1385; John Womack Jr., *Posición estratégica y fuerza obrera: Hacia una nueva historia de los movimientos obreros* (Fondo de Cultura Económica, 2008).

15. *Perth Amboy Evening News*, "Arrest 3 in Riot at AS&R Plant," January 10, 1917, 1; *Perth Amboy Evening News*, "Strike Action Here by N.J. Labor Heads," January 15, 1917, 1.

16. *The Patterson Morning Call*, "IWW Leaders Trailed," October 10, 1917, 8.

17. *Monthly Review of the Bureau of Labor Statistics* (Washington, DC: US Government Printing Office, 1918), 226; Annual Report Secretary of Labor, US Government Printing Office, 1919, 77.

18. *Perth Amboy Evening News*, "Roosevelt Will Oppose Reduction of Assessment," June 5, 1917, 5.

19. *Perth Amboy Evening News*, "Consider Salary Raise for Roosevelt Police," July 17, 1918, 5; *Perth Amboy Evening News*, "Denies U.S. Metals Refining Taken by Government," July 26, 1918, 2.

20. See the excellent work of John Womack on strategic position. John Womack Jr., "Technology, Work, and Strategic Positions in the Oil Industry in Mexico: 'Development,' 1908–1910," in *Memoria presentada en el Segundo Congreso Internacional de Historia Económica de la Asociación Mexicana de Historia Económica* (2004).

21. *San Francisco Chronicle*, "Strike Threatened at Kennett Mine," May 8, 1919, 10; *The Sacramento Star*, May 8, 1919, 2; *The Fresno Morning Republica*, "Deposits of Cadmium," May 9, 1919, 11.

22. *The Sacramento Bee*, "Mammoth Smelter Continues in Operation," May 12, 1919, 17; *The Sacramento Star*, "Strike at Mammoth Copper Co's Mine," May 13, 1919, 10; *The Sacramento Bee*, May 15, 1919, 9; *The Sacramento Bee*, "Smelter Still Running," May 15, 1919, 9; *The Sacramento Bee*, "Manager Says Smelter Will Be Closed," May 16, 1919, 8; *The Sacramento Bee*, "Operations at Smelter in Shasta End," May 17, 1919, 15; *San Francisco Chronicle*, "Heavy Shipments of Ore Arriving at Mammoth Smelter," May 19, 1919, 16.

23. *The Sacramento Bee*, "To Develop Keystone," May 21, 1919, 8; *San Francisco Chronicle*, "Mammoth Company Awaits Improved Copper Market," June 2, 1919, 16; *The Sacramento Bee*, "State Mineralogist to Aid in Appointment," July 16, 1919, 8; *The Sacramento Bee*, "Seek Reduction," July 12, 1919, 17; *The Sacramento Bee*, "Ask Lower Tax," July 15, 1919, 8; *The Sacramento Bee*, "Mammoth Copper Employees Resign," July 26, 1919, 17; *The Sacramento Bee*, July 28, 1919, 8; and "California Mining in 1919,"*EMJ*, vol. 109, 159.

24. *The Courier News*, "Chrome Workers Strike," September 4, 1919, 1; *EMJ*, vol. 108, October 11, 1919, 625.

25. AHCRMyP, FN, Contabilidad, libros auxiliares de registro, memorias de raya, file 2, p. 307; HBLL, USSRM Co. Records, 1901–71, box 7.

26. HBLL, USSRMCo Records 1901–71, box 3, folder 3; AHCRMyP, Relaciones

Laborales, Solicitudes de empleo, G1–G4. Even though the Cananea strike became a national symbol during the Mexican Revolution of 1910–20, the miners remained relatively marginal to the armed conflict. Mining operations were reduced during this decade, and mining companies were just starting to reach their optimal production levels. By the end of the conflict, some organizers could reach the unskilled workers in different camps. In May 1916, the Casa del Obrero Mundial at El Oro and Pachuca's mining camps organized a strike protesting the payment of wages with government bills, the *Infalsificables* printed by Carranza and that had quickly lost their value in wartime hyperinflation. The labor pause was relatively small, as only one plant of the firm operated on a reduced scale, but it was successful. The United States Company agreed to pay 50 percent of their wages in gold and sold subsidized corn to the workers. This first success of the strikers against the US Company was short-lived, however. It put more pressure on the local government than on the firm, and no workers' organization remained after the strike. Bernstein, *Mexican Mining Industry*, 103; ANON, "Conditions in Mexico," EMJ, CII (August 19, 1916): 340.

27. AHCRMyP, Fondo Norteamericano, Libros auxiliares de Registro, file 310; AHCRMyP, Correspondencia, Calland, vol. 12, file 33, fojas 989–992; AHCMRyP, Norteamericano, Beneficio y comercialización de metales, Superintendencia general de molinos, operaciones generales, vol. bajocama V, file 2, ff. 60–90.

28. Francisco Gorostiza, *Los ferrocarriles en la Revolución Mexicana* (México: Siglo XXI, 2013), 128; Juan Felipe Leal, *En la revolución (1910–1917)*, vol. 5 (México: Siglo XXI, 1988), 138–46.

29. USSRM, *Ax-I-Dent-Ax*, April 22, 1915, 15.

30. AHCRMyP, Fondo Norteamericano, Archivo Especial de la Dirección, Relaciones laborales, vol. 70, file 83 y vol. 132, file 4; AHCRMyP, Fondo Norteamericano, Correspondencia, Calland, vol. 12, file 33, p. 975 y file 35, fojas 230–41.

31. AHCRMyP, Fondo Norteamericano, Correspondencia, J. D. Smith, vol. 38, file 7, p. 61; Fondo Norteamericano, Correspondencia, General, file 77, page 101; FAPE-FCT, Archivo Plutarco Elías Calles, file 222, inventario 448, legajo ½; FAPEFCT, Archivo Fernando Torreblanca, Fondo Alvaro Obregón, fondo 11, serie 030500, file 1534, inventario 4406, legajo 1; AHCRMyP, Fondo Norteamericano, Correspondencia, Calland, vol. 12, file 33, p. 975 and file 35, fojas 230–41; AHCRMyP, Fondo Norteamericano, Correspondencia, J. D. Smith, vol. 38, file 7, p. 61; AHCRMyP, Fondo Norteamericano, Correspondencia, General, file 77, p. 101; AHCRMyP, Norteamericano Archivo especial de la Dirección, Administración, Oficina Pachuca, vol. 63, file 39, fojas 13–18; AHCRMyP, Norteamericano, Archivo especial de la Dirección, Administración, Oficina Pachuca, vol. 63, file 83.

32. AHCRMyP, Archivo Especial de la Dirección, Correspondencia, J. E. Smith, p. 91.

33. AHCRMYP, AED, Relaciones Laborales, File 83, vol. 70, file 4; vol. 132, file 6; vol. 123; AHCRMYP, AED, Administración, Oficina Pachuca, files 39 and 83, vol. 63; AHCRMyP, AED Reportes vol. 133, file 12; Correspondencia Calland, vol. 12, file 33 and file 35; Correspondencia J. D. Smith, file 7, vol. 38; AHCRMYP, FN,

Correspondencia General, file 77; FAPECT, Colección documental Embajada de Estados Unidos en México, inventario 29, legajo 1/6, file 080201, Series 1923; FAPECT, Archivo Plutarco Elías Calles, legajo ½, inventario 448, file 222; FAPECT, Fondo Álvaro Obregón, Archivo Fernando Torreblanca, legajo 1, inventario 4406, file 1534, series 030500, fondo 11; FAPECT, Archivo Plutarco Elías Calles Anexo, legajo 1/3, inventario 68, file 23, series 010400, fondo 12.

34. Juan Luis Sariego, *Enclaves y minerals en el norte de México: Historia social de los mineros de Cananea y Nueva Rosita, 1900–1970* (México, CIESAS, 1988), 196.

35. *EMJ*, vol. 94, no. 15, 676.

36. On June 1, 1906, over three thousand workers of the Cananea Copper Consolidated, in Sonora, started the first massive strike of the mining sector in Mexico in the twentieth century. The workers demanded a salary increase and the end of discrimination between Mexicans and white Americans, and, as an answer, the management of the American firm coordinated a brutal repression with armed company guards, the state government, and Arizona rangers. The authorities, institutional and de facto killed, apprehended, or purged in the next months the activists of the Partido Liberal Mexicano and the Western Federation of Miners (WFM). Nicolás Cárdenas García, "La huelga de Cananea en 1906: Una reinterpretación," *Estudios Sociológicos* (1998): 114–17: Michael J. Gonzales, "United States Copper Companies, the State, and Labour Conflict in Mexico, 1900–1910," *Journal of Latin American Studies* 26, no. 3 (1994): 651–81.

37. Lynn Robison Bailey, *Old Reliable: A History of Bingham Canyon, Utah* (Tucson, AZ: Westernlore Press, 1988), 77–98.

38. It was the time of the conflicts of Red Lopez and Joe Hill. Lynn Robinson Bailey, *The Search for Lopez: Utah's Greatest Manhunt* (Tucson, AZ: Westernlore Press, 1990, 3–15); and J. Kenneth Davies, "The Secularization of the Utah Labor Movement," *Utah Historical Quarterly* 45, no. 2 (1977): 108–34.

39. Skliris was a hiring agent for mining companies in Bingham, but also for the Western Pacific, Denver & Rio Grande, and the coal mines of Castle Gate, Hiawatha, Sunnyside, and Scofield. He also had offices in Denver and San Francisco. *Ogden Standard,* September 11, 1912, 3; *Salt Lake Tribune,* 1912: September 11, 1912, 1; September 12, 12; September 20, 1912, 1–2; September 22, 1; September 23, 2; September 24, 1, 11.

40. *Salt Lake Tribune,* 1912: May 3, 16; May 4, 10; May 5, 6; May 15, 2; May 17, 1–2; May 18, 1–2; May 21, 1; May 25, 6; September 11, 1; September 12, 12; September 20, 1–2; September 24, 1, 11; September 22, 1; September 23, 2. The *Coalville Times,* 1912: May 10, 2; May 24, 2. The *Ogden Standard,* September 11, 1912, 3. Allan Kent Powell, "A History of Labor Union Activity in the Eastern Utah Coal Fields: 1900–1934" (PhD diss., University of Utah, 1976), 120–49; Mellinger, *Race and Labor in Western Copper,* 5–6; John Ervin Brinley, "Western Federation of Miners" (PhD diss., University of Utah, 1972).

41. MLUU, Gaylon Hansen Papers, files 1–11, box 1.

42. MLUU, Gaylon Hansen Papers, files 1–11, box 1 and files 1–9, box 2.

43. *Bakerfield Californian,* April 12, 1924, 2; *Oakland Tribune,* April 13, 1924.

44. Womack, *Posición estratégica y fuerza obrera.*

45. *Fairbanks Daily News-Miner,* "Diesel Replacing Steam Nome Creek," April 8, 1930, 5; April 2, 1930, 8; March 31, 1930, 8; *Fairbanks Daily News-Miner,* "Alexander Is on Way to Outside," April 4, 1930, 8.

46. APR, Glenn Burrell is interviewed by Neville Jacobs in Fairbanks, Alaska, in September 1974, Oral History Archive, 76-04-01.

47. *Fairbanks Daily News-Miner,* "Open Cleary Road for Early Spring Traffic," April 3, 1930, 5; *Fairbanks Daily News-Miner,* "Chatanika Road Reported in Good Shape for Cars," April 15, 1930, 8; *Fairbanks Daily News-Miner,* "Legal Notice," January 15, 1930, 6.

48. Inspection on the site and conversation with Billie and Sari, ex-owner and owner of the Old FE Camp Hotel.

49. APR, Glenn Burrell is interviewed by Neville Jacobs in Fairbanks, Alaska, in September 1974, Oral History Archive, 76-04-01.

50. AHCRMyP, FN, Jurídico, Propiedad, series, contratos y denuncios, Tierras y aguas, box 1, file 5; AHCRMyP, FN, Contabilidad, libros auxiliares de registro, memorias de raya, file 3; AHCRMyP, FN, Jurídico, Propiedad, series, contratos y denuncios, Tierras y aguas, box 1, file 1; AHCRMyP, FN, Contabilidad, libros auxiliares de registro, Memorias de raya, file 307; FN, Jurídico, Propiedad, series, contratos y denuncios, Tierras y aguas, box 1, file 17.

51. MLUU, Special Collections, *Report of the Coal Inspector 1909–1916,* TN805.U8 A3. HBLL, USSRM Records, box 6, folder 7; HBLL, USSRM Records, box 5, folder 7; USSRM, *Ax-I-Dent-Ax,* March 1929, 4.

52. "The only favorable labor conditions which we have on the entire job are at this point, where the entire colony is composed of Finns. Apparently as long as there is no other nationality mixed in with them they get along very well." May 29, 1923, DD. Muir a N. W. Rice. MLUU, Gaylon Hansen papers, box 1, folder 7.

53. Scott Crump, *Bingham Canyon: The Richest Hole on Earth* (Salt Lake City: Great Mountain West Supply, 2005), 7–15.

54. MLUU, Gaylon Hansen papers, boxes 3, 8, 16; tube 40, extensions units 4&5.

55. Relaciones Laborales, solicitudes de empleo, vol. G1–G4.

56. MLUU, Gaylon W. Hansen papers, box 1; HBLL, USSRMCo. Records, 1901–71, box 3, folder 3; AHCRMyP, AED, Operación, Seguridad y Depto. Del Trabajo, vol. 73, file 99; MLUU, Gaylon Hansen papers, box 7, folder 10.

57. USSRM, *Ax-I-Dent-Ax,* July 1929, 13; January 1930, 26–29; January 1932, 25.

58. USSRM, *Ax-I-Dent-Ax,* October 1930, 9; October 1931, 29; September 1931, 37–38.

59. AHCRMyP, AED, Operación Seguridad y Depto. Del trabajo, vol. 72, file 96 y vol. 73, file 102; AHCRMyP, Correspondencia, General, vol. 48, file 48; MLUU, Gaylon Hansen papers, box 1, folder 4. These policies also exemplify the international influence of the Bureau of Mines regulations on firms. See Mark Aldrich, *Safety First:*

Technology, Labor, and Business in the Building of American Work Safety, 1870–1939, vol. 13 (Baltimore: Johns Hopkins Press, 1997), 13:211–58.

60. Only in the third district of Utah, the company faced forty-four cases between 1903 and 1936, of workers' accidents in their location. See the cases State of Utah Third District Court, 4728, 5310, 5328, 5544, 6249, 6665, 7012, 7794, 8358, 9632, 9993, 21165, 22613, 23996, 24310, 24362, 25531, 26818, 28848, 29016, 29221, 29581, 29907. See a similar trajectory on the effects on the body of workers in coal mines in Mark Aldrich, "Preventing 'The Needless Peril of the Coal Mine': The Bureau of Mines and the Campaign against Coal Mine Explosions, 1910–1940," *Technology and Culture* 36, no. 3 (1995): 483–518.

61. AHCRMyP, Operación Seguridad y Departamento del trabajo, vol. 73, exp. 101.

CHAPTER 3. COSMOPOLITAN DISEASES

1. Mines Register, vol. 14, American Metal Market Company, 1918, 1650–1651. Poor's Manual of Industrials; Manufacturing, Mining and Miscellaneous Companies, 1918, 965.

2. MITDC, Waldemar Lindgren Papers, MC 419 b2, folder "Mining Reports from Hidalgo, Mexico."

3. Jolande Jacobi Paracelsus and Norbert Guterman, *Paracelsus: Selected Writings* (New York: Pantheon Books, 1951), 94–121.

4. Abraham Gottlob Werner, *New Theory of the Formation of Veins: With Its Application to the Art of Working Mines* (London: A. Constable, 1809); James Hutton, *Theory of the Earth: With Proofs and Illustrations*, vol. 1 (Library of Alexandria, 1795); Charles Lyell, "On the Structure and Probable Age of the Coal-Field of the James River, near Richmond, Virginia," *Quarterly Journal of the Geological Society* 3, nos. 1–2 (1847): 261–80; Charles Lyell, "On Foot-Marks Discovered in the Coal-Measures of Pennsylvania," *Quarterly Journal of the Geological Society* 2, nos. 1–2 (1846): 417–20; Charles Lyell, "On Certain Trains of Erratic Blocks on the Western Borders of Massachusetts, United States" (London: Proceedings at the Meeting of the Members of the Royal Institution of Great Britain, 1855), 86–97; and Charles Lyell, "On the Relative Age and Position of the So-Called Nummulite Limestone of Alabama," *Quarterly Journal of the Geological Society* 4, nos. 1–2 (1848): 10–17.

5. See the excellent work of Lucier on the formation of consulting during the emergence of states and companies in the nineteenth century. Paul Lucier, *Scientists and Swindlers: Consulting on Coal and Oil in America, 1820—1890* (Baltimore: Johns Hopkins University Press, 2008).

6. Waldemar Lindgren, *Mineral Deposits* (New York: McGraw-Hill Book Company, Incorporated, 1913); Waldemar Lindgren, "The Copper Deposits of the Clifton-Morenci District, Arizona" (Washington, DC: US Government Printing Office, 1906); Waldemar Lindgren, "The Gold Belt of the Blue Mountains of Oregon: US Government Printing Office," in *The Gold Belt of the Blue Mountains of Oregon: Twenty-Second*

Annual Report of the United States Geological Survey to the Secretary of the Interior 1901 (Washington, DC: Government Printing Office), 1902, 754–55.

7. Transactions AIME, 1915, xvi; AHCRMyP, Correspondencia, M. H. Kuryla, vol. 28, file 73 and 81; MIT, *Geology at* MIT, *1865–1965*, vol. I, 385–461.

8. MITDC, William Otis Crosby papers, MC 68, box 14.

9. Waldemar Lindgren and Clyde P. Ross, "The Iron Deposits of Daiquiri, Cuba," *Transactions of the American Institute of Mining Engineers* 53 (1916): 40–66; Waldemar Lindgren, "The Tin Deposits of Chacaltaya, Bolivia," *Economic Geology* 19, no. 3 (1924): 223–28; and Waldemar Lindgren and Edson Sunderland Bastin, "The Geology of the Braden Mine, Rancagua, Chile," *Economic Geology* 17, no. 2 (1922): 75–99.

10. MITDC, Waldemar Lindgren Papers, MC419, b2, folder 4.

11. Waldemar Lindgren, *Mineral Deposits* (New York: McGraw-Hill Book Company, =Incorporated, 1913), 2.

12. Bulletin of the MIT, president's report, 1917.

13. John W. Servos, "The Industrial Relations of Science: Chemical Engineering at MIT, 1900–1939," *Isis* 71, no. 4 (1980): 531–49.

14. William H. Walker, "The Spirit of Alchemy in Modern Industry," *Science* 33, no. 859 (1911): 913–18.

15. Bulletin of the MIT, president's report, 1917, 69; Christophe Lécuyer, "MIT, Progressive Reform, and Industrial Service,' 1890–1920," *Historical Studies in the Physical and Biological Sciences* 26, no. 1 (1995): 35–88.

16. Bulletin of the MIT, president's report, 1917, 69; Lécuyer, "MIT, Progressive Reform."

17. Ortega Morel, "Minería y tecnología," 140–45; Michael H. Kuryla and Galen Clevenger, "H. Liquid-Oxygen Explosives at Pachuca," *Transactions of American Institute of Mining and Metallurgical Engineers* 69 (1923), 321–40.

18. National Research Council, *Industrial Research Laboratories of the United States* (New York: R. R. Bowker Company, 1920), 80, 101.

19. In 1952 the Kennecott transferred their central research center to the University of Utah. *Electrical Engineering Journal* vol. 41 (July 1922): 150; *Great Falls Tribune*, March 15, 1942, 10; The *Montana Standard,* June 29, 1947, 16; USSRM, *Ax-I-Dent-Ax*, November 1929, 1; National Research Council, *Industrial Research Laboratories*.

20. MIT, Bulletin of the MIT, president's report, 1919–20.

21. MIT, Bulletin of the MIT, president's report, 1918, 49.

22. USSRMCo, *Ax-I-Dent-Ax*, 1929–32; Herbert Megraw, *The Flotation Process* (New York: McGraw-Hill Book Company Inc.); Bernstein, *Mexican Mining Industry*, 138; Ortega Morel, "Minería y tecnología," 138–46; A. D. Akin, "Practical Manufacture," *Engineering and Mining Journal* (1923): 978–80; AHCRMyP, Operación, minas, vol. 66, file 1, MIT, 1917.

23. W. C. Smith and A. A. Heimrod, "Application of the Cottrell Process to the Recovery of Fume from Silver Refinery Operations," *Chem &Met Engineering* 21 (1919): 360.

24. Bulletin of the MIT, president's report, 1920, 51.

25. William H. Walker, "The Technology Plan," *Science* 51, no. 1319 (1920): 357–59; and David F. Noble, *America by Design: Science, Technology, and the Rise of Corporate Capitalism* (New York: Oxford University Press, 1979), 142–44.

26. "The Triumph and Tragedy of the Technology Endowment Campaign," *Chemical Age* vol. 28 (1933): 341–43.

27. Bulletin of the MIT, president's report, 1917–22, Leo P. Brophy, "Origins of the Chemical Corps"; Brooks E. Kleber and Dale Birdsell, *The Chemical Warfare Service: Chemicals in Combat* (Washington, DC: Office of the Chief of Military History, US Army, 1966), 3–36; and Leo P. Brophy and George J. B. Fisher, *The Chemical Warfare Service: Organizing for War*, vol. 1. (Washington, DC: Office of the Chief of Military History, Department of the Army, 1959), 3–17.

28. *Iron Age* vol. 104 (1919): 492.

29. MITDC, AC 13 b21, folder "U.S. Army-Chemical Warfare Section."

30. USSRM Reports, 1916.

31. AHCRMyP, Relaciones Laborales, vol. 123, file 8; *Mining Science*, February 11, 1909, 113; and *Mining Science*, May 1909, 32.

32. Jordan N. Burns, Rudofo Acuna-Soto, and David W. Stahle, "Drought and Epidemic Typhus, Central Mexico, 1655–1918," *Emerging Infectious Diseases* 20, no. 3 (2014): 442; and Martha Eugenia Rodríguez, "El tifo en la Ciudad de México en 1915," *Gaceta médica de México* 152, no. 2 (2016): 253–58.

33. CHMHU, Richard Strong papers, GA 82, b 24, folder 49.

34. "The Nobel Prize in Physiology or Medicine 1928," NobelPrize.org, accessed June 16, 2022, https://www.nobelprize.org/prizes/medicine/1928/nicolle/lecture/; Kim Pelis, *Charles Nicolle, Pasteur's Imperial Missionary: Typhus and Tunisia* (Rochester, NY: University of Rochester Press, 2006), 91–97.

35. William C. Summers, *The Great Manchurian Plague of 1910–1911: The Geopolitics of an Epidemic Disease* (New Haven, CT: Yale University Press, 2012), 86–87.

36. George Shattuck, "Work in Serbia," in *The Harvard Volunteers in Europe: Personal Records of Experience in Military, Ambulance, and Hospital Service*, ed. Mark Antony De Wolfe Howe (Cambridge, MA: Harvard University Press, 1916), 67.

37. CHMHU, Richard Strong papers, GA 82, box 24, folder 49.

38. CHMHU, Richard Strong papers, GA 82, b2, folder 3; b 13, folder 40.

39. Richard P. Strong, E. E. Tyzzer, Charles T. Brues, A. W. Sellards, and J. C. Gastiaburu, "Verruga peruviana, Oroya fever and uta: Preliminary report of the first expedition to South America from the Department of Tropical Medicine of Harvard University," *Journal of the American Medical Association* 61, no. 19 (1913): 1713–16; Marcos Cueto, "Tropical Medicine and Bacteriology in Boston and Peru: Studies of Carrion's Disease in the Early Twentieth Century," *Medical History* 40, no. 3 (1996): 344–64; and David Salinas-Flores, "One Hundred Years after the Expedition by Harvard University to Peru to Investigate Carrion's Disease: Lessons for Science," *Revista de la Facultad de Medicinam*, 64, no. 3 (2016): 517–24.

40. That was the case of the Duquesne Mining and Reduction Company when hiring Dr. Ames. CHMHU, Richard Strong papers, GA82, b 19, folder 85; and the Standard Oil Co. in hiring Dario Gutierrez. CHMHU, Richard Strong papers, GA82, b2 0, f 48; or the doctors in Liberia of Firestone, CHMHU, Richard Strong papers, GA82, b 19, folder 117.

41. Eli Chemin, "Richard Pearson Strong and the Iatrogenic Plague Disaster in Bilibid Prison, Manila, 1906," *Reviews of Infectious Diseases* 11, no. 6 (1989): 996–1004; Kristine A. Campbell, "Knots in the Fabric: Richard Pearson Strong and the Bilibid Prison Vaccine Trials, 1905–1906," *Bulletin of the History of Medicine* 68, no. 4 (1994): 600–638.

42. Richard P. Strong and B. C. Crowell, "The Etiology of Beriberi," *Philippine Journal of Science* 7, no. 4 (1912).

43. Victor G. Heiser, "American Sanitation in the Philippines and Its Influence on the Orient," *Proceedings of the American Philosophical Society* 57, no. 1 (1918): 60–68; V. G. Heiser, "Recollections of Cholera at the Turn of the Century," *Bulletin of the New York Academy of Medicine* 47, no. 10 (1971); Warwick Anderson, "Modern Sentinel and Colonial Microcosm: Science, Discipline, and Distress at the Philippine General Hospital," *Philippine Studies* (2009): 153–77; and Aaron Rom O. Moralina, "State, Society, and Sickness: Tuberculosis Control in the American Philippines, 1910–1918," *Philippine Studies* 2009, 179–218.

44. Victor G. Heiser, "Leprosy in the Philippine Islands," *Public Health Reports* (1896–1970), August 13, 1909.

45. Victor G. Heiser, "Reminiscences on Early Tropical Medicine," *Bulletin of the New York Academy of Medicine* 44, no. 6 (1968): 654–60; Summers, *Great Manchurian Plague of 1910–1911.*

46. Anne-Emanuelle Birn, *Marriage of Convenience: Rockefeller International Health and Revolutionary Mexico* (Rochester, NY: University of Rochester Press, 2006).

47. Rocio Gomez, *Silver Veins, Dusty Lungs: Mining, Water, and Public Health in Zacatecas, 1835–1946* (Lincoln: University of Nebraska Press, 2020), 171.

48. Henry Pomeroy Davison, *The American Red Cross in the Great War*, vol. 46 (New York: Macmillan, 1919).

49. Armando Solorzano, "Sowing the Seeds of Neo-Imperialism: The Rockefeller Foundation's Yellow Fever Campaign in Mexico," *International Journal of Health Services* 22, no. 3 (1992): 529–54.

50. SLHU, Papers of Alice Hamilton, A-22, Vt-34, box 2.

51. Matthew Ringenberg and Joseph Brain, *The Education of Alice Hamilton: From Fort Wayne to Harvard* (Bloomington: Indiana University Press, 2019).

52. Alice Hamilton, *Exploring the Dangerous Trades—The Autobiography of Alice Hamilton, MD* (Redditch, UK: Read Books Ltd., 2013), 73–95.

53. In 1902, she produced a report on the typhoid epidemic that had started that summer in the area and attributed the spread of the disease to the action of flies. The theory, simple and reinforcing the mainstream, won rapid scientific acclaim, although Hamilton soon discovered that the reason of the spread was the poor conditions of the

drainage systems around the marginal areas of the city. This disease, caused by the *Salmonella Typhi* bacteria usually spread through contaminated water and food, should not be confused to typhus fever described in the previous section, which is caused by the *Rickettsia Prowazekii* bacteria carried by lice. Alice Hamilton, "The Fly as a Carrier of Typhoid: An Inquiry into the Part Played by the Common House Fly in the Recent Epidemic of Typhoid Fever in Chicago," *Journal of the American Medical Association* 40, no. 9 (1903): 576–83.

54. Jane Addams, *Twenty Years at Hull-House* (Urbana: University of Illinois Press, 1990), 297–202.

55. Hamilton, *Exploring the Dangerous Trades,* 114–26.

56. Hamilton, *Exploring the Dangerous Trades,* 128.

57. Alice Hamilton, *Nineteen Years in the Poisonous Trades* (New York: Harper & Bros., 1929).

58. Alice Hamilton, *Lead Poisoning in the Smelting and Refining of Lead,* Bulletin of the US Bureau of Labor Statistics (Washington, DC: US Government Printing Office, 1914).

59. Hamilton, *Nineteen Years in the Poisonous Trades,* 583.

60. Hamilton, *Nineteen Years in the Poisonous Trades,* 30–46; HBLL, box 16, folder 4.

61. Alice Hamilton, *Industrial Poisons in the U.S.* (New York: The Macmillan Co., 1929), 19–109; Alice Hamilton and Harriet L. Hardy, "Industrial Toxicology" (New York: Harper and Brothers Publishing, 1974), 20–63.

62. Hamilton, *Nineteen Years in the Poisonous Trades,* 584.

63. Hamilton, *Nineteen Years in the Poisonous Trades,* 324.

64. SLHU, papers of Alice Hamilton, A-22, Vt-34, box 1, folder 5.

65. Periódico oficial del Estado de Hidalgo, February 16, 1919, 2.

66. Periódico Oficial del Estado de San Luis Potosí, December 4, 1918, 8; Ryan M. Alexander, "The Spanish Flu and the Sanitary Dictatorship: Mexico's Response to the 1918 Influenza Pandemic," *The Americas* 76, no. 3 (2019): 443–65; Periódico Oficial del Estado de Sinaloa, November 1, 1918, 5; Periódico Oficial del Estado de Yucatan, December 7, 1918, 3.

67. Ryan M. Alexander, "The Spanish Flu and the Sanitary Dictatorship: Mexico's Response to the 1918 Influenza Pandemic," *The Americas* 76, no. 3 (2019), 447; Lourdes Márquez Morfín and América Molina del Villar, "El Otoño de 1918: Las Repercusiones de La Pandemia de Gripe En La Ciudad de México," *Desacatos* no. 32 (2010): 121–44; Miguel Ángel Cuenya, "México ante la pandemia de influenza de 1918: Encuentros y desencuentros en torno a una política sanitaria," *Astrolabio* 13 (2014).

68. Periódico oficial del Estado de Hidalgo, November 1, 1918, 1.

69. Periódico oficial del Estado de Hidalgo, November 24, 1918, 2; November 8, 1918, 1; April 16, 1919, 2; Alexander, "Spanish Flu and the Sanitary Dictatorship."

70. AHCRMyP, FN, Correspondencia, D.S. Calland, vol. 11, folder 26.

71. *The Salt Lake Tribune,* November 20, 1918, 11; November 19, 1918, 11.

72. EMJ, v. 106, 1918, 1131.

73. EMJ, v. 107, 1919, 56.

74. MLUU, Gaylon Hansen, box 2, folder 4.

75. In 1925 the young law student Miguel Aleman Valdes made some field work around the sick by silicosis in the Real del Monte area, making a dissertation titled "Disease and Professional Risks."

76. Theophrastus Paracelsus, *Four Treatises of Theophrastus von Hohenheim Called Paracelsus* (Baltimore: Johns Hopkins University Press, 1996), 58. See the studies on the origins of medical geology in Brian E. Davies et al., "Medical Geology: Perspectives and Prospects," in *Essentials of Medical Geology* (Dordrecht: Springer, 2013), 1–13.

77. Jolande Jacobi Paracelsus, and Norbert Guterman, *Paracelsus: Selected Writings* (New York: Pantheon Books, 1951), 17.

78. Rocio Gomez finds a similar dynamic in the relationship between the body of miners and the mineral bodies in Zacatecas in the late nineteenth century. See Gomez, *Silver Veins, Dusty Lungs*, 113–50.

79. AHCRMyP, FN AED, Seguridad y depto del trabajo, file 102.

80. AHCRMyP, FN AED, Seguridad y depto del trabajo, file 102.

81. Tomas G. Perrin, "Contribución al examen histopatológico de la silicosis pulmonar en Mexico: Nota primera algunas consideraciones sobre cien exámenes microscópicos," *Departamento del Trabajo*, 1934; Tomas G. Perrin "Contribución al estudio histopatológico de la silicosis pulmonar en México," *Gaceta Médica de México*, November 8, 1933.

82. Oscar A. Glaeser, "Ventilation of Small Metal Mines and Prospect Openings," *Transactions of the American Institute of Mining, Metallurgical and Petroleum Engineers* 126 (1937): 221; Oscar Arthur Glaeser, "Intermittent Mine Ventilation at the United Verde Mine as an Economy Measure" (PhD diss., University of Washington, 1932); O. A. Glaeser, "Gold Bullion Mill and Cyanide Plant, Willow Creek, Alaska," *Mining & Scientific Press* 126 (1921): 601–5.

83. AHCRMyP, FN AED, Seguridad y depto del trabajo, files 100 and 102.

84. AHCRMyP, FN Relaciones Laborales, vol. 123, exp 8.

85. Panama Canal Zone Governor, Annual Report, 1916.

86. MITDC, Warren Mead Papers, MC 625 b1, *Economic Geology of the Iron Ore deposits of the Anshan District in South Manchuria*, 1921; Yoshihisa Tak Matsusaka, *The Making of Japanese Manchuria, 1904–1932* (Leiden: Brill, 2020).

87. Panama Canal Zone Governor, Annual Report, 1937; Robert R. Shrock, "WJ Mead, Experimental Geologist," *Science* 132, no. 3435 (1960): 1235–36; US Department of Interior, The Hoover Dam Documents, US Government Printing Office, 1948; US Colorado River Board, Report on the Boulder Dam Project, 1928.

88. Warren Mead, "Report on Geological Studies in the Pachuca—Real Del Monte Silver District," MITDC, *Warren Mead Papers* MC (n.d.): 625 b2; AHCRMyP, FN Relaciones Laborales, vol. 123, exp. 8.

89. Mead, "Report on Geological Studies in the Pachuca—Real del Monte Silver District," MITDC, Warren Mead Papers, MC 625 b2.

90. The head of the department of MIT disagreed, in the discussion, with the dating of the faulting in the horizontal veins of Pachuca made by Edward Wisser, two years before after several spending a considerable amount of time in the area. Wisser had identified a small period of mineralization in the at the end of the extrusion period, meaning the end of the more active volcanic activity in the area. In contrast, Mead established that mineralization occurred significantly earlier, more intermittently, and in a larger amount of time, before the intrusion of significant amounts of quartz. Another piece of the debate between the experts was dating the faulting. Wisser had dated the displacements of the horizontal veins in Pachuca as premineral, meaning that the fault was created first and then became mineralized by magma activity. Mead, in contrast, believed that he had experimental evidence than the faults were postmineral, meaning that tectonic activity produced displacements on the vein after they charged it with minerals. Edward Hollister Wisser, "Some Applications of Structural Geology to Mining in the Pachuca-Real Del Monte Area, Pachuca Silver District, Mexico," *Economic Geology* 41, no. 1 (1946): 77–86.

91. MITDC, MIT Office of the President, AC4, b 169, folder 16; AC 4, b 128, f 8; AC 4, b 13, "American; Smelting"; AC4 b226, "U.S. Smelting Company."

92. Strong to Frederic N. Watriss, International Petroleum Company, June 6, 1928. CHMHU, Richard Strong papers, GA82, b20, f. 127.

93. CHMHU, Richard Strong papers, GA82, b19, folder 134; GA82, b19, f117; Richard P. Strong et al., *The African Republic of Liberia and the Belgian Congo, Based on the Observations Made and Material Collected during the Harvard African Expedition 1926–1927* (New York: Greenwood Press Publishers, 1930).

94. CHMHU, Joseph C. Aub Papers, MC30, b7, folder 16.

95. Joseph Aub to Harriet Hardy, December 11, 1946. CHMHU, Joseph C. Aub Papers, MC30, b7, folder 3.

96. USSRM Report, 1942–50.

97. Oscar A. Glaeser, "Compensation for Lung Changes Due to the Inhalation of Silica Dust," *Archives of Industrial Hygiene and Occupational Medicine* 2 (1950): 56.

98. In 1950, the US Steel Company asked J. R. Killian, as president of MIT, to nominate arbitrators into the contract negotiation with the United Steelworkers of America. MITDC, MIT Office of the President, AC4, box 226, folder 20.

CHAPTER 4. SOVIET OF TECHNICIANS

1. MITDC, AIME, *Boston Section Minutes*, 129.

2. Thorstein Veblen, *The Engineers and the Price System*, vol. 31 (New Brunswick, NJ: Transaction Publishers, 1921).

3. John Hays Hammond, "Russia of Yesterday and Tomorrow," *Scribner's Magazine* LXXI (May 1922): 515–27; Mark Hendrickson, "The Sesame That Opens the Door of Trade: John Hays Hammond and Foreign Direct Investment in Mining, 1880–1920," *The Journal of the Gilded Age and Progressive Era* 16, no. 3 (2017): 325.

4. F. W. Draper, "The Story of the Russian Platinum Shipment of 1917," *Mineral Industry* XXVI (1918): 550–55.

5. Alan Knight, *The Mexican Revolution: Counter-Revolution and Reconstruction*, vol. 2 (Lincoln: University of Nebraska Press, 1990), 374; USSRMCo, *Annual Report*, 1914; AHCRMP, FN, JD, Actas e informes, vol. 1, file 2; ACHRMyP, Norteamericano, Dirección General, Correspondencia D. S. Calland, vol. 9, file 13. 12-18-1914; AHCRMyP, Otras Compañías, vol. 119, folder 111.

6. This overview draws from the author's examination of the USSRMCo, Annual Reports, 1906–52; and Kennecott, Kennecott Copper Corporation Annual Reports, 1895–98, 1916–33, 1942–60.

7. AHCRMyP, Op. Seguridad y Departamento del trabajo, vol. 73. file 98. AHCRMyP, AED, Administración Oficina Boston, vol. 59, file 16.

8. Charles V. Mutschler, *Wired for Success: The Butte, Anaconda & Pacific Railway, 1892–1985* (Pullman, WA: Washington State University, 2002).

9. Robert L. Spude, *Science and Technology in Alaska's Past* (Anchorage, AK: Alaska Historical Society, 1990), 187–94; National Park Service, "Kennecott Story," National Park Service, US Department of the Interior, https://www.nps.gov/wrst/learn/historyculture/upload/Kennecottbulletin.pdf; Marcosson, *Metal Magic*, 91; and US Congress, *Government of Alaska: Statements before the Committee on Territories, United States Senate, on the Bill S. 5436* (Washington, DC: Government Printing Office, 1910).

10. Marcosson, *Anaconda*, 205. The control of freight was central, as well, in the conformation of the United Fruit Company. See Colby, *Business of Empire*.

11. USF Co., *Thirty Years of Coal*; USSRMCo, *Annual Report*, 1912; Curtis Seltzer, *Fire in the Hole: Miners and Managers in the American Coal Industry* (Lexington: University Press of Kentucky, 1985), 37; and USSRMCo, *Annual Report*, 1918–30; USSRMCo, *Ax-I-Dent-Ax*, January 1930, 2.

12. USSRMCo, *Ax-I-Dent-Ax*, March 1919, 23; *Ax-I-Dent-Ax*, September 1931, 15; *Ax-I-Dent-Ax*, June 1929, 1–11; *Ax-I-Dent-Ax*, January 1930, 29; and *Ax-I-Dent-Ax*, September 1930, 6.

13. *The Magazine of Wall Street and Business Analyst*, vol. 32, 52.

14. AHMCRMyP, Administración oficina Boston, vol. 58, folder 9; AHCMRyP, Relaciones Laborales, vol. 123, files 8, 15, 30.

15. See the studies of Stephen J. Kobrin, "Expatriate Reduction and Strategic Control in American Multinational Corporations," *Human Resource Management* 27, no. 1 (1988): 63–75; Yasmin Merali, "Individual and Collective Congruence in the Knowledge Management Process," *The Journal of Strategic Information Systems* 9, nos. 2–3 (2000): 213–34; Hong Chung, Patrick T. Gibbons Lai, and Herbert P. Schoch, "The Management of Information and Managers in Subsidiaries of Multinational Corporations," *British Journal of Management* 17, no. 2 (2006): 153–65.

16. "This spirit of the body predominates equally in all its parts; by which the members must consider themselves as having a common interest, and that common

interest separated both from that of the country which sent them out and from that of the country in which they act." Burke, *Works of the Right Honorable Edmund Burke,* 354.

17. USSRMCo, *Ax-I-Dent-Ax,* July 1930, 10; September 1930, 16; December 1930, 27; ACHRMyP, Correspondencia, Calland, vol. 7, file 1.

18. AHCRMyP, Archivo Especial de la Dirección, vol. 74, file 114.

19. Robert W. Randall, "British Company and Mexican Community: The English at Real Del Monte, 1824–1849," *The Business History Review* (1985): 622–44.

20. William Schell, *Integral Outsiders: The American Colony in Mexico City, 1876–1911* (Lanham, MD: Rowman & Littlefield, 2001), ix–29.

21. *Handbook of Travel,* prepared by the Harvard Travelers Club (Cambridge, MA: Harvard University Press, 1917).

22. USSRMCo, *Ax-I-Dent-Ax,* June 1930, 1.

23. In December 1930 the officials of the US Company in Mexico regretted the transfer of the Lantz family. After being employed as general manager of the Santa Gertrudis Company for years, Mr. Lantz had to leave due to the economic difficulties of the company after the crash. USSRMCo, *Ax-I-Dent-Ax,* June 1930, 4; December 1930, 27; September 1930, 16.

24. AHCRMyP, FN, Operación. Seguridad y Departamento del trabajo. vol. 73, file 98; AHCRMyP, FN, Jurídico, Propiedad, series, contratos y denuncios, Tierras y aguas, box 1, file 17.

25. USSRMCo, *Ax-I-Dent-Ax,* May 15, 1932.

26. USSRMCo, *Ax-I-Dent-Ax,* June 1930, 4; January 1930.

27. For tropical masculinity and technical identity, see James W. Martin, *Banana Cowboys: The United Fruit Company and the Culture of Corporate Colonialism* (Albuquerque: University of New Mexico Press), 2018.

28. USSRMCo, *Ax-I-Dent-Ax,* January 1930, 27; May 1930, 20; November 1931.

29. USSRMCo, *Ax-I-Dent-Ax,* May 1930, 20; October 1931; September 1931, 16; APR, Oral History Archive, Harry Badger et al. interview in Fairbanks, Alaska in 1955, 02-00-135-03.

30. USSRMCo, *Ax-I-Dent-Ax,* August 1931, 34; September 1931; January 1930, 26; December 1930, 31.

31. USSRMCo, *Ax-I-Dent-Ax,* July 1930, 8; March 1929, 21; April 1929, 20; January 1930, 30.

32. USSRMCo, *Ax-I-Dent-Ax,* March 1929; September 1930, 9; July 1929.

33. USSRMCo, *Ax-I-Dent-Ax,* September 1930, 9; September 1930, 17; September 1931, 34; October 1930, 18.

34. USSRMCo, *Ax-I-Dent-Ax,* May 1930, 18, 32; AHCRMyP, AED, Misceláneos, vol. 132, file 25.

35. AHCRMyP, Exploración, vol. 106, file 90.

36. USSRMCo, *Ax-I-Dent-Ax,* November 1929, 1; March 1931; June 1932; National Research Council (US), *Industrial Research Laboratories;* Ernest Patty, "Farthest-North

College," ALASKA PER F901 P2; AHCRMyP, Correspondencia, M. H. Kuryla, vol. 27, files 30 and 54.

37. Bernstein, *Mexican Mining Industry*, 156.

38. AHCRMYP, Relaciones Laborales, vol. 123, file 8, pp. 16–20.

39. Beatty, *Technology and the Search*, 181–207.

40. The diffusion of cyanidation and flotation also replaced another old technician's trade formation: the assayer. See Robert Lester Spude, *To Test by Fire: The Assayer in the American Mining West, 1848–1920* (Urbana: University of Illinois Press, 1989), 136–51.

41. USSRMCo, *Ax-I-Dent-Ax*, December 1930, 14.

42. Brian Leech, "Competition, Community, and Entertainment: The Anaconda Company's Promotion of Mine Safety in Butte, Montana, 1915–1942," *The Mining History Journal* 24, no. 1 (2017): 19–39.

43. Marcosson, *Metal Magic*, 264; *Safety Engineering*, vol. 41, 1921, 32.

44. USSRMCo, *Firing Line*.

45. USSRMCo, *Ax-I-Dent-Ax*, October 1931, 14.

46. USSRMCo, *Ax-I-Dent-Ax*, December 1930, 14.

47. Clark C. Spence, *Mining Engineers and the American West: The Lace-Boot Brigade, 1849–1933* (New Haven, CT: Yale University Press, 1970), 5–15, 286–294.

48. USSRMCo, *Ax-I-Dent-Ax*, February 1931, 17; April 1931, 25; April 1921, 1–4; May 1932, 6, 43; January 1932, 7–8; January 1931, 1–9; July 1932, 12; June 1929, 16–18; July 1930, 4.

49. USSRMCo, *Ax-I-Dent-Ax*, May 1932, 35.

50. For an exploration on the relationship between the Western frontier ideals and the American expansion to the tropics, see Martin, *Banana Cowboys*.

51. USSRMCo, *Ax-I-Dent-Ax*, July 1929, 15; October 1930, 10; October 1929, 5, 51; May 1930, 34; April 1931; September 1931, 16; Report on Mosquito Investigations, USSRM, 1931.

52. USSRMCo, *Ax-I-Dent-Ax*, March 1929, 24; September 1931, 16; March 1931, 44–45; October 1932, 9; AED, Misceláneos, vol. 132, exp. 25.

53. E. James Dixon, *Bones, Boats & Bison: Archeology and the First Colonization of Western North America* (Albuquerque: University of New Mexico Press, 1999), 55–57; "University of Alaska Museum of the North," Tumblr, *University of Alaska Museum of the North* (blog), January 29, 2015, https://alaskamuseum.tumblr.com/post/109507993040/effie-the-wooly-mammoth-on-a-late-summer-day.

54. USSRMCo, *Ax-I-Dent-Ax*, January–July 1932; June 1930, 4.

55. USSRMCo, *Ax-I-Dent-Ax*, June 1929, 1–11.

56. John C. Boswell, "History of Alaskan Operations of United States Smelting," *Refining, and Mining Company. Mineral Industries Research Laboratory, University of Alaska, Fairbanks* (1979), 65.

57. ALMP, boxes 11–12; Alan Probert, "Bartolomé de Medina: The Patio Process and the Sixteenth Century Silver Crisis," in *Mines of Silver and Gold in the Americas* (New York: Routledge, 2020), 96–130.

58. Alan Probert, "En pos de la plata" (Pachuca: UAEH), 2011, 446.

59. AIME, *Transactions*, New York, 1920, xxi.

60. AIME, *Transactions*, New York, 1920, xxvii.

61. AIME, *Transactions*, New York, 1918, lvii–lix; AIME, Bi-monthly Bulletin of the American Institute of Mining Engineers (Princeton, NJ: Princeton University), 1919, xlviii–lxiii; John Hillman, "The Emergence of the Tin Industry in Bolivia," *Journal of Latin American Studies* 16, no. 2 (1984): 403–37; Manuel E. Contreras, "The Bolivian Tin Mining Industry in the First Half of the Twentieth Century," ISA *Research Papers* 32 (1993).

62. Edward S. Kaplan, *US Imperialism in Latin America: Bryan's Challenges and Contributions, 1900–1920*, no. 35 (Westport, CT: Greenwood Publishing Group, 1998), 61–62.

63. USSRMCo, *Ax-I-Dent-Ax*, December 1930, 32; February 1931, 4; May 1930, 7–8; HBLL, USSRMCo Records, 1901–71, box 2.

64. USSRMCo, *Ax-I-Dent-Ax*, December 1932, 32. *The Michigan Technic*, vol. 44, March 1931, 16–17; *The Michigan Alumnus*, vol. 37, 217.

65. "Russian Soviet Wants Private Capital," EMJ, 116, 19123, no. 2, 74.

66. Philip S. Gillette, "American-Soviet Trade in Perspective," *Current History (Pre-1986)* 65, no. 000386 (1973): 158.

67. MITDC, AC 69, b 10 (geology).

68. "American Engineers to Inspect Soviet Union Coal Mines," *Economic Review of the Soviet Union* II, no. 7, 2; Henry J. Freyn, "Impressions and Experiences in the USSR," *The Society of Industrial Engineers Bulletin* (1931): 12–22.

69. Henry J. Freyn, "Russian Industry: Can America Afford to Neglect It?" *Class and Industrial Marketing* (January 1931): 42–43, 56–58.

70. Freyn, "Impressions and Experiences," 12–22.

71. USSRMCo, *Ax-I-Dent-Ax*, May 1932, 22; July 1932, 17; October 1932, 9; December 1930, 32; Walter Scott Dunn, *The Soviet Economy and the Red Army, 1930–1945* (Westport, CT: Greenwood Publishing Group, 1995), 12; and Wilson T. Bell, *Stalin's Gulag at War: Forced Labour, Mass Death, and Soviet Victory in the Second World War* (Toronto: University of Toronto Press, 2018), 73.

72. USSRMCo, *Ax-I-Dent-Ax*, January 1932, 8.

73. *Foreign Minerals Quarterly: A Regional Review of Foreign Mineral Resources, Production, and Trade*, supplement to vol. 1, no. 2, Washington DC, April 1938, 20; USSR Gosplan, "The Soviet Union Looks Ahead: The Five-Year Plan for Economic Construction," (New York: Liveright, 1929), 50–58, 273–75.

CHAPTER 5. AN AUTOPSY IN EVERY CASE

Some paragraphs of this chapter had a previous version in García Solares, "Striking Hard Rock Veins. Multinational Corporations and Miners' Unions in Mexico and the United States, 1906–1952."

1. USSRM, *Annual Reports*, 1915–1940; *Copper, Curb and Mining Outlook,* January 10,

1917, 21; *Poor's Manual of Industrials, Manufacturing, Mining, and Miscellaneous Companies,* New York: 1916, vol. 7, 1245.

2. Hearing before a Subcommittee of the Committee on Education and Labor, US Senate, vol. 9, July 9, 1940, 62–63; *Stop Silicosis* (US Department of Labor, 1938); David Rosner and Gerald Markowitz, *Deadly Dust: Silicosis and the Politics of Occupational Disease in Twentieth-Century America* (Princeton, NJ: Princeton University Press, 1994), 200.

3. Frances Perkins, "What You Really Want Is an Autopsy: Opening Remarks of Frances Perkins to the Tri State Silicosis Conference in Joplin, Missouri, 1940," in *Milestone Documents of American Leaders: Exploring the Primary Sources of Notable Americans,* ed. Paul Finkelman (Dallas: Schlager Group, 2009), 1691.

4. Alice Hamilton, "A Mid-American Tragedy," *Sur-Vey Graphic* 29 (1940): 434–37.

5. SLHU, Alice Hamilton Papers, A22, box 1, folders, 24, 19, 16, 14, 13; box 5, p. 83; Norah Hamilton, "Scenes from Soviet Russia," *Fortnightly* 126 (1929): 171–79.

6. NLRB, *Case No. R-2001- Decided September 21, 1940,* vol. 27, Decisions and Orders of the NLRB (Washington, DC: US Government Printing Office, 1942), 383–88.

7. Bernstein, *Mexican Mining Industry,* 156; file 9, vol. 123, Relaciones Laborales, AED, AHCRMyP; file 39, vol. 26, Correspondencia, Kuryla, AHCRMyP; Federico Besserer, José Díaz, and Raúl Santana, "Formación y consolidación del sindicalismo minero en Cananea," *Revista Mexicana de Sociología* (1980): 1321–53.

8. AHCRMyP, Correspondencia, Kuryla, file 13, vol. 25; file 39, vol. 26 and file 56, vol. 28; AHCRMyP, AED, Relaciones Laborales, files 13–14, vol. 123.

9. Andres Hijar, "Where Is Our Revolution?: Workers in Ciudad Juarez and Parral-Santa Barbara during the 1930s" (PhD diss., Northern Illinois University, 2015), 119–51.

10. AHCRMyP, AED, Relaciones Laborales, vol. 123, files 5 and 13; Victoria Novelo, "De huelgas, movilizaciones y otras acciones de los mineros del carbón de Coahuila," *Revista Mexicana de Sociología* (1980): 1358–59; Luis Emilio Cacho Jiménez, "La fundación del Sindicato Minero-Metalúrgico," in *Los sindicatos nacionales en el México contemporáneo, Minero-Metalúrgico,* ed. Javier Aguilar (México: GV Editores, 1987), 7–38; Juan Luis Sariego, *Enclaves y minerales en el norte de México: Historia social de los mineros de Cananea y Nueva Rosita, 1900–1970* (México: CIESAS, 1988), 207–11; Luis Reygadas, *Proceso de trabajo y acción obrera: historia sindical de los mineros de Nueva Rosita, 1929–1979* (México: INAH, 1988), 56; and Rossana Cassigoli, *Napoleón Gómez Sada: Liderazgo sindical y cultura minera en México* (México: Miguel Ángel Porrúa, 2004), 109–24.

11. MLUU, Gaylon Hansen papers, files 4–8, box 6, and folders 7–11, box 7; HBLL, USSRM Co. Records, 1901–71, file 14, box 13; Robert H. Zieger, *The CIO, 1935–1955* (Chapel Hill: University of North Carolina Press, 1997), 22–42; *Ogden Standard-Examiner,* October 2, 1936, 2; October 10, 1936,–2; October 11, 1936, 1–2; *The Daily Herald,* October 7, 1936, 2; October 8, 1936, 2; October 11, 1936, 1–2; October 11, 1936, 1.

12. *The Salt Lake Tribune,* October 24, 1936, 10; October 18, 1936, 36; October 22, 1936, 2; November 3, 1936, 5; November 11, 1936, 12; October 10, 1936, 17; November 23,

1936, 8; December 4, 1936, 15; December 10, 1936, 1, 9; December 7, 1936, 10; December 8, 1936, 1, 4; December 12, 1936, 10; December 15, 1936, 1, 4; December 19, 1936, 6, 36; December 12, 1936, 10; December 15, 1936, 1, 4; December 19, 1936, 6, 36.

13. UCB-IUMMSW, file 26, box 38; folder "#444 Fairbanks," box 131.

14. Charles Wyzanski, "Brief for the USSRM," LAW HDSC, paige box 001.

15. Decisions and orders of the NLRB, vol. 27, pp. 383–386.

16. UCB-IUMMSW, file 26, box 38; folder "#444 Fairbanks," box 131; file 26, box 28; file 67, box 43; file 24, box 80.

17. UCB-IUMMSW, folder "#444," box 173.

18. HBLL, USSRM Records, box 3, file 2.

19. UCB-IUMMSW, box 38, folder 26.

20. AHCRMyP, AED, Relaciones Laborales, vol. 126, file 48.

21. UCB-IUMMSW, Convention Proceedings, 1946, p. 70.

22. AHCRMyP, AED, Relaciones Laborales, files 5 and 11, vol. 123 and file 43 vol. 126; *El Informador*, 18 de Mayo de 1936, 3.

23. HBLL, file 21, box 13, USSRM Co. Records, 1901–71; *Salt Lake Tribune*, May 22, 1946, 13; June 12, 1946, 1.

24. AHCRMyP, AED, Relaciones Laborales, file 43, vol. 126.

25. USSRM reports, 1942/1947, Mineral Yearbook, 1947, p. 1486.

26. UCB-IUMMSW, box 125, folder "Minutes of USSRM Council"; box 199, folder 15; box 278, folder "Mexican Miners Union."

27. *Periódico Oficial del Estado de Chihuahua*, 1944-06-19, p. 2122.

28. AHEH, Fondo Sindicato, Secretaría del Interior, Exterior y Actas, Actas 1944, legajo 11.

29. National Mining Congress, *Report of the Proceedings of the National Mining Congress: Held in the People's Theatre, in the City of Denver, Colorado, on... Nov. 18, 19 and 20, 1891* (Denver: News Print. Co., 1892).

30. AHCRMyP, AED, Relaciones Laborales, pp. 12, file 48, vol. 126.

31. Congress of Industrial Organizations (US) and James B. Carey, *Report of the CIO Delegation to the Soviet Union* (Washington, DC: Congress of Industrial Organizations, 1945); UCB-IUMMSW, "Official Proceedings," 1946; George E. Lichtblau, "The World Federation of Trade Unions," *Social Research* 25, no. 1 (1958): 1–36; and Jon V. Kofas, "US Foreign Policy and the World Federation of Trade Unions, 1944–1948," *Diplomatic History* 26, no. 1 (2002): 21–60.

32. Luis Sariego, *El Estado y la minería mexicana: Política, trabajo y sociedad durante el siglo XX* (SEMIP, 1988), 230–45.

33. AHCRMyP, AED, Relaciones Laborales, file 48, vol. 126.

34. UCB-IUMMSW, box 122, folder "Mexican Miners Union"; file 26, box 38; UCB-IUMMSW, "Convention proceedings"; AHEH, SITMRM, Actas 1947, Secretaria del Interior, Exterior y Actas.

35. UCB-IUMMSW, Official Proceedings, 42nd Convention, Cleveland, Ohio, 1946.

36. James J. Palomino Lorence, *Clinton Jencks and Mexican-American Unionism in the American Southwest* (Urbana: University of Illinois Press, 2013), 41–48.

37. Lorence, *Clinton Jencks,* 87.

38. HBLL, USSRM Records, box 10, folder 3.

39. Mario T. García, *Mexican Americans: Leadership, Ideology, and Identity, 1930–1960,* vol. 36 (New Haven, CT: Yale University Press, 1989), 201.

40. UCB-IUMMSW, box 6, folder 17; UCB-IUMMSW, box 57, folder 46.

41. Lorence, *Clinton Jencks,* 87; Ellen R. Baker, *On Strike and on Film: Mexican American Families and Blacklisted Filmmakers in Cold War America* (Chapel Hill: UNC Press Books, 2007), 72.

42. US Bureau of International Labor, Directory of the WTFU, p. 42, 1958.

43. UCB-IUMMSW, box 120, folder "Mexican Miners"; box 123, folder "Mexican Miners Union."

44. Sariego, *El Estado y la minería mexicana,* 239–44; Novelo, "De huelgas, movilizaciones."

45. Lorence, *Clinton Jencks,* 100–149.

46. Harvey Matusow, *False Witness* (New York: Cameron & Kahn, 1955).

47. Eric Pace, "Charles E. Wyzanski, 80, Is Dead," *New York Times,* section A, September 5, 1986; Josiah Bartlett Lambert, *"If the Workers Took a Notion": The Right to Strike and American Political Development* (Ithaca, NY: Cornell University Press, 2005), 167–72; Archibald Cox papers, LAW HDSC, box 59, folder 6.

48. Archibald Cox and John T. Dunlop, "Regulation of Collective Bargaining by the National Labor Relations Board," *Harvard Law Review* 63, no. 3 (1950): 389–432; Archibald Cox and John T. Dunlop, "The Duty to Bargain Collectively During the Term of an Existing Agreement," *Harvard Law Review* 63, no. 7 (1950): 1097–1133.

49. Staughton Lynd, "Government without Rights: The Labor Law Vision of Archibald Cox," *Industrial Relations Law Journal* 4, no. 3 (1981): 483–95.

50. Charles E. Wyzanski, "Labor Disputes during the War: An Address Given at Northwestern University, January 12, 1942," LAW.

51. Archibald Cox, *Law and the National Labor Policy* (Los Angeles: Institute of Industrial Relations, University of California, 1960), 81.

52. Archibald Cox, "Some Aspects of the Labor Management Relations Act, 1947," *Harvard Law Review* 61, no. 1 (1947): 1–49.

53. Archibald Cox, "Revision of the Taft-Hartley Act," *West Virginia Law Review* 55 (1952): 100.

54. Charles E. Wyzanski, "The Open Window and the Open Door: An Inquiry into Freedom of Association," *California Law Review* (1947): 336–51.

55. *Judge Wyzanski's Report of His Trip to Africa.* Confidential, September 1960. LAW, Comp, 964 WYZ.

56. Barry Castleman, "Alice Hamilton and the FBI," *International Journal of Occupational and Environmental Health* 22, no. 2 (April 2016): 173–74, https://doi.org/10.1080/10773525.2016.1156922; Alice Hamilton Documents, FBI Documents/FOIA, WVGE

04–0004603, available at https://www.ncbi.nlm.nih.gov/pmc/articles/PMC4984968/
bin/YJOH_A_1156922_SM5774.pdf.

CHAPTER 6. SPACE OF EXCEPTION

1. H. V. Winchell, "Geology of Pachuca and El Oro, Mexico," *Transactions of the American Institute of Mining Engineers* 66 (1921): 27–40.

2. MLUU, Gaylon Hansen papers, tubes 38–40; HBLL, USSRMCo Records, 1901–71, box 11, folder 12; box 12, folder 1.

3. R. R. Bryan and M. H. Kuryla, "Milling and Cyanidation at Pachuca," *Transactions of the American Institute of Mining Engineers* 112 (n.d.): 722–33.

4. Hobbes, *Leviathan*, 117–18.

5. Carl Schmitt, *Political Theology: Four Chapters on the Concept of Sovereignty* (Chicago: University of Chicago Press, 2005), 5–15; Giorgio Agamben, "State of Exception," in *State of Exception* (Chicago: University of Chicago Press, 2021), 1–31.

6. "The decision on the exception is a decision in the true sense of the word. Because a general norm, as represented by an ordinary legal prescription, can never encompass a total exception, the decision that a real exception exists cannot therefore be entirely derived from this norm." Schmitt, *Political Theology*, 6.

7. Carl Schmitt, *The Nomos of the Earth* (New York: Telos Press, 2003), 98–99.

8. Ronald H. Coase, "The Problem of Social Cost," *The Journal of Law and Economics* 3 (1960): 1–44.

9. Joseph Smith, *The Book of Mormon: An Account Written by the Hand of Mormon upon Plates Taken from the Plates of Nephi*, vol. 2 (Salt Lake City: The Church of Jesus Christ of Latter-day Saints, 1921), 2 Ne. 5:19–24.

10. Howard A. Christy, "Open Hand and Mailed Fist: Mormon-Indian Relations in Utah, 1847–52," *Utah Historical Quarterly* 46, no. 3 (1978): 219.

11. "And he leadeth away the righteous into precious lands, and the wicked he destroyed, and cursed the land unto them for their sakes." Joseph Smith, *The Book of Mormon: An Account Written by the Hand of Mormon upon Plates Taken from the Plates of Nephi*, vol. 2 (Salt Lake City: The Church of Jesus Christ of Latter-day Saints, 1921), 1 Ne. 17:38.

12. Christy, "Open Hand and Mailed Fist," 226.

13. W. P. Reeve, *Making Space on the Western Frontier: Mormons, Miners, and Southern Paiutes* (Urbana: University of Illinois Press, 2010), 11–35; J. Lindell, "Mormons and Native Americans in the Antebellum West" (master's thesis, San Diego State University, 2011); T. G. Alexander, "Irrigating the Mormon Heartland: The Operation of the Irrigation Companies in Wasatch Oasis Communities, 1847–1880," *Agricultural History* 76, no. 2 (2002): 172–87; L. J. Arrington, *Great Basin Kingdom: An Economic History of the Latter-day Saints, 1830–1900* (Urbana: University of Illinois Press, 2005), 220–42.

14. HBLL, Correspondence of the Benjamin Winchester Family, folder 1.

15. M. A. Church, "Smoke Farming: Smelting and Agricultural Reform in Utah, 1900–1945," *Utah Historical Quarterly* 72, no. 3 Summer (2004): 198.

16. SLCC, Complaints 5909 and 6120.

17. SLCC, Complaints 6116, 7128, 7173, 7200, 7201, 7257, 7258, 7259, 7267, 7286, 7288, 7319, 7408, 7682, 7736, 7774, 9840.

18. HBLL, USSRMCo Records, 1901–71, box 5, folder 2; Edward Thompson, *American and English Annotated Complaints, Containing the Important Complaints Selected from the Current American, Canadian, and English Reports*, Bancroft-Whitney, 1916, 12; and David Gerard and Timothy J. LeCain, "Property Rights and Technological Innovation," in *The Technology of Property Rights*, ed. Terry J. Anderson and Peter J. Hill (Lanham, MD: Rowman and Littlefield Publishers, 2001), 147–78.

19. EMJ, August 7, 1909, vol. 88, p. 280; January 8, 1910, vol. 89, p. 91; *The Mining Magazine*, vol. 1, 1909, p. 296; *The Chemical Age*, vol. X, no. 1909, p. 146; MLUU, boxes 1–6; HBLL, USSRMCo Records, 1901–71, box 1, folder 9; box 16, folder 4.

20. HBLL, USSRMCo Records, 1901–71, box 1, folder 1; box 2, folder 9.

21. Gonzalo Amozurrutia, "El Concepto de indio y el trabajo minero en Latinoamérica en la época colonial: La Rebelión de Real del Monte en 1766" (master's diss., Facultad de Filosofía y Letras, Universidad Nacional Autónoma de México, 2017), 96–146.

22. I. M. P. Moreno. "Coacción y disensión: Protestas frente a los repartimientos mineros en Perú y Nueva España, siglo XVIII," *Estudios de historia novohispana* (2015): 1–17.

23. C. R. Ruiz Medrano, "El tumulto de abril de 1757 en Actopan: Coerción laboral y las formas de movilización y resistencia social de las comunidades indígenas," *Estudios de Historia Novohispana*, no. 36 (2007): 108–9.

24. AHCRMyP, FN, Jurídico, Propiedad, series, contratos y denuncios, Tierras y aguas, box 1, file 9; AGN, Instituciones Coloniales, General de Parte 051, vol. 31, file 311, p. 225; vol. 67, file 454, pp. 255–56; Probert, "En pos de la plata," 214–20.

25. V. M. Ballesteros, *Síntesis de la Guerra de Independencia en el estado de Hidalgo*, vol. 1 (Pachuca: UAEH, 2005).

26. Rocío Ruiz Barrera, "Breve historia de Hidalgo" (USA: Fondo De Cultura Economica, 2000), xviii; AHA, Aguas Nacionales, box 345, file 3697, legajo 01; box 663, file 7551, legajo 2/2.

27. R. Huizar Álvarez, "Carta hidrogeológica de la cuenca del Río de las Avenidas, de Pachuca, Hgo., México," *Investigaciones geográficas* 27 (1993): 95–131; AHCRMyP, Fondo Norteamericano, Propiedad, series, contratos y denuncios, Tierras y aguas, box 1, file 5.

28. Gobierno San Salvador, "Reclamo a Morrill B. Spaulding" (San Salvador: Imprenta Nacional, 1903).

29. AHCRMyP, Fondo Norteamericano, Jurídico, Propiedad, series, contratos y denuncios, Tierras y aguas, box 1, file 2; Junta Directiva, Actas e informes, vol. 1, file 2.

30. AHCRMyP, Fondo Norteamericano, Jurídico, Propiedad, series, contratos y denuncios, Tierras y aguas, box 1, file 2.

31. AHA, Fondo Aprovechamientos superficiales. box 252, file 6067.

32. Periódico Oficial del Estado de Hidalgo, 12/08/1912.

33. Periódico Oficial del Estado de Hidalgo, 07/16/1912; 08/08/1912; 12/24/1912; 12/28/1912.

34. Ortega Morel, "Minería y tecnología," 210–25.

35. Aha, Fondo Aprovechamientos superficiales, box 252, file 6067, pp. 11–15.

36. L. Ferleger, "Uplifting American Agriculture: Experiment Station Scientists and the Office of Experiment Stations in the Early Years after the Hatch Act," *Agricultural History* 64, no. 2 (1990): 5–23; V. E. Rice, "The Arizona Agricultural Experiment Station: A History to 1917," *Arizona and the West* 20, no. 2 (1978): 123–40; and Joseph Horne Jeppson, *The Secularization of the University of Utah, to 1920* (Berkeley: University of California Press, 1973), 103–8.

37. D. H. Dyal, "Mormon Pursuit of the Agrarian Ideal," *Agricultural History* 63, no. 4 (1989): 19–35.

38. Jeppson, *Secularization of the University*, 103–8; HBLL, James E. Talmage Papers, vol. XI, personal journal, ff. 40–78; US Congress, Reed Smoot, a senator from the state of Utah, to hold his seat (January 16, 1904–April 13, 1906).

39. B. Kime, "Exhibiting Theology: James E. Talmage and Mormon Public Relations, 1915–20," *Journal of Mormon History* 40, no. 1 (2014): 208–38.

40. J. A. Widtsoe, *The Relation of Smelter Smoke to Utah Agriculture* (Logan, UT: Experiment Station of the Agricultural College of Utah, 1903), 178. One year before, Widtsoe and Merrill had published a report on the poisoning of cattle by the consumption of beet pulp contaminated with lead. J. A. Widtsoe and L. A. Merrill, "Lead Ore in Sugar Beet Pulp," *Bulletin of the Utah Agricultural Experiment Station*, no. 74 (1902).

41. John Andreas Widtsoe, *Rational Theology: As Taught by the Church of Jesus Christ of Latter-day Saints* (Salt Lake City: Deseret News, 1915); C. D. Ford, "Materialism and Mormonism: The Early Twentieth-Century Philosophy of Dr. John A. Widtsoe," *Journal of Mormon History* 36, no. 3 (2010): 1–26; "It is a familiar fact that Joseph Smith taught that matter is eternal and has not been nor can be created. Matter is coexistent with God. God himself is material in the sense that His body is composed of a refined kind of matter." J. A. Widtsoe, "Joseph Smith as Scientist: A Contribution to Mormon Philosophy," Salt Lake City: General Board, 1908, 148.

42. HBLL, USSRMCo Records, 1901–71, box 2, folder 8.

43. HBLL, USSRMCo Records, 1901–71, box 2, folder 8.

44. HBLL, USSRMCo Records, 1901–71, box 1, folder 1; box 6, folder 1; Church, "Smoke Farming," 204.

45. HBLL, USSRMCo Records, 1901–71, box 1, folder 1; USSRMCo, *Ax-I-Dent-Ax*, July 1929, 21–22; USSRMCo, *Annual Report*, 1911–21; Church, "Smoke Farming," 215.

46. Complaints: 9435, 9923, 10272, 11601, 12231, 17167, 17736, SLCC.

47. "The defendant Anaconda Copper Mining Company agrees that it will at all times use its best efforts to prevent, minimize and ultimately to completely eliminate the emission and distribution from its smelting works at Anaconda, Mont., of

all deleterious fumes, particularly those containing sulfur dioxide," quoted in Fredric Lincoln Quivik, *Smoke and Tailings: An Environmental History of Copper Smelting Technologies in Montana, 1880–1930* (Philadelphia: University of Pennsylvania Press, 1998), 348. See also Donald MacMillan, *Smoke Wars: Anaconda Copper, Montana Air Pollution, and the Courts, 1890–1924* (Helena: Montana Historical Society, 2000), 167–88.

48. *Salt Lake Tribune*, January 8, 1914, 12; HBLL, USSRMCo Records, 1901–71, box 16, folder 3; MLUU, George James Peirce papers, 1914–15, "Report on the Injury to Vegetation due to Smelter Smoke in the Salt Lake Valley, Utah in 1915."

49. HBLL, USSRMCo Records, 1901–71, box 1, folder 7; USSRMCo, *Annual Report*, 1921.

50. HBLL, USSRMCo Records, 1901–71, box 1, folder 7; box 15, folder 8; USSRM, *Annual Report*, 1920–21.

51. MLUU, Gaylon Hansen papers, box 8, folder 6.

52. A. Ochoa Serrano, *Los agraristas de Atacheo* (Zamora: El Colegio de Michoacán AC, 1989), 86–92.

53. J. H. Mogica, *Organización campesina y lucha agraria en el Estado de Hidalgo, 1917–1940* (Pachuca: UAEH, 2000), 55–93; Venustiano Carranza, Ley Agraria del 6 de enero de 1915, available at http://www.pa.gob.mx/publica/rev_58/analisis/leypercent20agrariapercent20delpercent 206percent20depercent20Januarypercent20depercent201915.pdf.

54. AHA, Aprovechamientos superficiales, box 252, file 6067.

55. AHCRMyP, Fondo Norteamericano, Junta Directiva, Actas e informes, vol. 1, file 2; Correspondencia, D. S. Calland, vol. 07, file 02, pp. 492–496.

56. AHCRMyP, Correspondencia, D. S. Calland, vol. 23, file 33, p. 935.

57. AHCRMyP, Fondo Norteamericano, Correspondencia, D. S. Calland, vol. 23, file 33; Correspondencia, M. H. Kuryla, vol. 25, file 13; AHA, Aprovechamientos superficiales, box 3268, file 52968.

58. AHA, Aprovechamientos Superficiales, box 863, file 12392; AHCRMyP, Fondo Norteamericano, Correspondencia, D. S. Calland, vol. 12, file 33.

59. AHA, Aprovechamientos superficiales. box 252, file 6067.

60. AHA, Aprovechamientos superficiales, box 252, file 6067.

61. Kuryla, "Milling and Cyanidation"; AHCRMyP, Fondo Norteamericano, Archivo Especial de la Dirección, Relaciones Laborales, vol. 123, file 11; AHCRMyP, Correspondencia, M. H. Kuryla, vol. 25, file 13.

62. AHA, Aprovechamientos Superficiales, box 3591, file 49793; Aguas Nacionales, box 345, file 3697, legajo 01; Aguas nacionales box 663, file 7551, legajo 2/2; USSRMCo, *Annual Report*, 1926–30.

63. John D. Rees, "Effects of the Eruption of Parícutin Volcano on Landforms, Vegetation, and Human Occupancy," *Volcanic Activity and Human Ecology* (1979):249–92.

64. J. M. M. Arroyo, *Historia y narrativa en el ejido de San Francisco Uruapan (1916–1997)* (Zamora: El Colegio de Michoacán AC, 2002), 79.

65. ALMP, box 20.

66. Rees, "Effects of the Eruption"; Martha Gabriela Gómez-Vasconcelos, "El volcán Paricutín en el campo volcánico Michoacán-Guanajuato: una revisión," *Ciencia Nicolaita* 74 (2018): 15–30.

67. ALMP, box 20.

68. AhCrMyp, FN, correspondencia, general, vol. 56, folder 182.

69. HBLL, USSRMCo Records, 1901–71, box 11, folder 12; G. F. Knowlton, A. P. Sturtevant, and C. J. Sorenson, "Adult Honeybee Losses in Utah as Related to Arsenic Poisoning," *Utah State Univ., Agric. File Stn., Bull.* (1950), 340; T. C. Yao and G. F. Knowlton, "Surface Arsenic Occurrence on Some Plants Attractive to Bees," *Utah Agr. File Sta. Mimeo. Ser, 349,* 1–8. See a modern assessment of the effects of arsenic in soils in G. Morin and G. Calas, "Arsenic in Soils, Mine Tailings, and Former Industrial Sites," *Elements* 2, no. 2 (April 1, 2006): 97–101, https://doi.org/10.2113/gselements.2.2.97.

70. US Census of Population and Housing, 1880, West Jordan Precinct; US Census of Population and Housing, 1940, Midvale Precincts, available at ancestry.org.

71. Censo General de la República Mexicana, Dirección General de Estadística, 1895; 6° Censo de Población de los Estados Unidos Mexicanos, Secretaria de la Economía Nacional, 1940.

72. AHA, Aprovechamientos Superficiales, box 359, file 49793; Aguas Nacionales, box 345, file 3697, legajo 01; Aguas nacionales box 663, file 7551, legajo 2/2. USSRMCo, *Annual Report,* 1926–30; Hernández, *Organización campesina,* 53; Ruiz, "Breve historia de Hidalgo."

EPILOGUE

1. AHEH, Fondo Sindicato, Secretaría del Interior, Exterior y Actas, Actas 1947, S. 146.

2. The consumption of the mineral reserves as a share of the income in the Mexican subsidiary was 14.92 percent. This was below the 27 percent for the rest of the holding, the 27.99 percent in the Western division, the 31.35 percent of Alaska's properties, and the 91.89 percent of the coal division. "USSRM Co, Quarterly Directors Meeting, December 23, 1938," HBLL, USSRMCo Records, 1901–71, box 4, folders 5–6.

3. USSRMCo, *Annual Reports,* 193548; AHCRMyP, FN, JD, Actas e Informes, vol. 4, file 20; José Alfredo Uribe Salas, "Historia económica y social de la compañía y cooperativa minera 'Las Dos Estrellas' en El Oro y Tlalpujahua, 1898–1959," Universidad Michoacana de San Nicolás de Hidalgo y Consejo Superior de Investigaciones Científicas de España (CSIC), Morelia (Michoacán México) (2010), 497.

4. USSRMCo, *Annual Report,* 1950–60.

5. USSRMCo, *Annual Report,* 1946–75.

6. United States et al., *Multinational Corporations and United States Foreign Policy: Hearings, Ninety-third Congress [Ninety-fourth Congress, Second Session],* vol. 4 (Washington, DC: US Government Printing Office, 1973), 517.

7. ALMP, box 22, folder 18.

8. E. Carrington Pardo, "Saneamiento general en relación con las enfermedades de

origin hídrico en San Miguel Regla, Hgo" (PhD diss., México, Universidad Nacional Autónoma de México, 1947).

9. Por: Pachuca VIVE, "Estadio Hidalgo cumple 29 años; así fue su construcción e inauguración," February 14, 2022, https://www.pachucavive.com/web/estadio-hidalgo -cumple-29-anos-asi-fue-su-construccion-e-inauguracion/; "El Estadio Hidalgo Celebra Su 25 Aniversario," ESTO (blog), February 14, 2018, https://www.esto.com.mx/ 335321-noticias-pachuca-aniversario-estadio-hidalgo-hoy-se-cumplen-25-anos-de-la -inauguracion-del-estadio-hidalgo/; and Eleazar Salinas-Rodríguez et al., "Leaching of Silver Contained in Mining Tailings, Using Sodium Thiosulfate: A Kinetic Study," *Hydrometallurgy* 160 (March 1, 2016): 6–11, https://doi.org/10.1016/j.hydromet.2015.12 .001.

10. Angela Caputo, "East Chicago Residents Fleeing Lead Contamination Find Few Housing Options," *The Chicago Tribune,* October 5, 2016; Michael Hawthorne, "7 Years Later, New Study Shows East Chicago Kids Exposed to More Lead Because of Flawed Government Report," *The Chicago Tribune,* August 20, 2018. See the environmental struggles of the area in Andrew Hurley, *Environmental Inequalities: Class, Race, and Industrial Pollution in Gary, Indiana, 1945–1980* (Chapel Hill: University of North Carolina Press, 1995). For a heritage-centered approach to dealing with mine tailings, see Sean M. Gohman, "It's Not Time to Be Wasted: Identifying, Evaluating, and Appreciating Mine Wastes in Michigan's Copper Country," *IA: The Journal of the Society for Industrial Archeology* (2013): 5–22.

11. Tony Semerad, "Zion's Bank Breaks Ground on Midvale Campus," *Salt Lake Tribune,* August 19, 2020. The ranchers around the Anaconda Copper Company continued to experience the effects of the tailings of the corporation, again, in coordination with the conservationist approach of the National Parks System. See a discussion in Fredric L. Quivik, "Landscapes as Industrial Artifacts: Lessons from Environmental History," *IA: The Journal of the Society for Industrial Archeology* 26, no. 2 (2000): 55–64.

Bibliography

ARCHIVES

AGN (Archivo General de la Nación)

AHA (Archivo Histórico del Agua)

AHCRMyP (Archivo Histórico de la Compañía de Real del Monte y Pachuca)

AHEH (Archivo Histórico del Estado de Hidalgo)

ALMP (Alan & Lillie M. Probert Collection, Benson Latin American Collection, The University of Texas at Austin)

APR (Alaska & Polar Regions Collections & Archives, Rasmuson Library, University of Alaska Fairbanks)

BCY (Brown Collection, Yale Library)

CHMHU (Center for the History of Medicine, Harvard University)

CEHM (Centro de Estudios Históricos Carso)

EMJ (*Engineering and Mining Journal*)

FAPECFT (Fideicomiso Archivos Plutarco Elías Calles y Fernando Torreblanca)

HBLL (Harold B. Lee Library Special Collections, Brigham Young University)

LAW (Special Collections, Library of the Harvard School of Law)

MITDC (MIT Distinctive Collections)

MLUU (Marriot Library Special Collections, University of Utah)

SLCC (Salt Lake County Court Records)

SLCR (Salt Lake County Recorder Office)

SLHU (Schlesinger Library, Harvard University)

UCB-IUMMSW (University of Colorado Boulder, Western Federation of Miners/ International Union of Mine, Mill, and Smelter Workers collection)

USHS (Utah State Historical Society)

LEGAL CASES

Santa Clara County v. Southern Pacific Railroad Company, 118 U.S. 394 (1886)

Mike Erceg vs. Fairbanks Exploration Company, 95 F.2d 850 (9th Cir. 1938)

Godfrey vs. American Smelting Refinery Co., 158 F. Rep. 227 (D. Utah 1907)

John F. Cowan et al. vs. U.S. Mining Co., Complaint No. 3297, Reel 88 (3rd District Court, Salt Lake County, 1900)

U.S. Mining Co. vs. Thomas Pells, Complaint No. 3351, Reel 90 (3rd District Court, Salt Lake County, 1900)

U.S. Mining Co. vs. Thomas Pells, Complaint No. 3352, Reel 90 (3rd District Court, Salt Lake County, 1900)

Thomas Pells vs. U.S. Mining Company, Complaint No. 3480, Reel 93 (3rd District Court, Salt Lake County, 1903)

U.S. Mining vs. St. Joe Mining Co., Complaint No. 4356, HBLL (3rd District Court, Salt Lake County, 1903)

Giovani Lavagnino vs. U.S. Mining Company, Complaint No. 5333, Reel 143 (3rd District Court, Salt Lake County, 1903)

S. M. Whitmore vs. U.S. Smelting Co and Bingham Copper & Gold Mining Company, Complaint No. 5909, Reel 151 (3rd District Court, Salt Lake County, 1903)

Jesse Argent vs. U.S. Smelting Co., Complaint No. 6116, Reel 155 (3rd District Court, Salt Lake County, 1904)

S. M. Whitmore vs. U.S. Smelting Co. and Bingham Copper & Gold Mining Company, Complaint No. 6120, Reel 155 (3rd District Court, Salt Lake County, 1906)

Moylan C. Pox vs. U.S. Mining Co., Complaint No. 6699, Reel 165 (3rd District Court, Salt Lake County, 1904)

Michael Foley et al. vs. U.S. Mining Co., Complaint No. 6926, Reel 168 (3rd District Court, Salt Lake County, 1905)

James Lane vs. U.S. Smelting Co., Complaint No. 7128, Reel 171 (3rd District Court, Salt Lake County, 1905)

Frank L. Plam and Charlotte vs. Utah Consolidated Mining Company, Complaint No. 7173, Reel 172 (3rd District Court, Salt Lake County, 1905)

Robert Elwood vs. U.S. Smelting Co., Complaint No. 7257, Reel 173 (3rd District Court, Salt Lake County, 1905)

W. H. Bird vs. U.S. Smelting Co., Complaint No. 7258, Reel 173 (3rd District Court, Salt Lake County, 1905)

Andrew Thomson vs. U.S. Smelting Co., Complaint No. 7259, Reel 173 (3rd District Court, Salt Lake County, 1905)

Alma Hogenson and Johanna Hogenson vs. U.S. Smelting Co., Complaint No. 7286, Reel 173 (3rd District Court, Salt Lake County, 1905)

Emmanuel F. Lennberg and Christena Lennberg vs. U.S. Smelting Co., Complaint No. 7288, Reel 173 (3rd District Court, Salt Lake County, 1905)

Benjamin Winchester vs. U.S. Smelting Co., Complaint No. 7319, Reel 174 (3rd District Court, Salt Lake County, 1905)

Mahala Higgins and Robert Ellwood vs. U.S. Smelting Co., Complaint No. 7408, Reel 176 (3rd District Court, Salt Lake County, 1905)

Michael Foley et al. vs. Niagara Mining & Smelting Company, Complaint No. 7676, Reel 178 (3rd District Court, Salt Lake County, 1905)

Ira Beakstead vs. U.S. Smelting Co., Complaint No. 7682, Reel 179 (3rd District Court, Salt Lake County, 1905)

Michael Foley et al. vs. U.S. Mining Co., Complaint No. 7714, Reel 179 (3rd District Court, Salt Lake County, 1905)

Michael Foley et al. vs. U.S. Mining Co., Complaint No. 7715, Reel 179 (3rd District Court, Salt Lake County, 1905)

William L. Turner vs. U.S. Smelting Co., Complaint No. 7736, Reel 180 (3rd District Court, Salt Lake County, 1905)

Carl Jensen vs. U.S. Smelting Co., Complaint No. 7774, Reel 180 (3rd District Court, Salt Lake County, 1905)

Henry H. Turner vs. U.S. Smelting Co., Complaint No. 7876, Reel 182 (3rd District Court, Salt Lake County, 1906)

Con Jones vs. U.S. Smelting Co., Complaint No. 7932, Reel 183 (3rd District Court, Salt Lake County, 1906)

Charles M. Hansen vs. U.S. Smelting Co., Complaint No. 8225, Reel 187 (3rd District Court, Salt Lake County, 1906)

George Holmbergh vs. U.S. Smelting Co., Complaint No. 8583, Reel 193 (3rd District Court, Salt Lake County, 1906)

Alma Hogenson vs. U.S. Smelting Co., Complaint No. 8593, Reel 193 (3rd District Court, Salt Lake County, 1906)

Michael McMillan vs. U.S. Smelting and Refining Co., Complaint No. 9435, Reel 204 (3rd District Court, Salt Lake County, 1907)

Ira Bennion vs. U.S. Smelting Co., Complaint No. 9840, Reel 211 (3rd District Court, Salt Lake County, 1907)

Neil Anderson vs. U.S. Smelting Co., Complaint No. 9923, Reel 212 (3rd District Court, Salt Lake County, 1908)

John Anderson vs. U.S. Smelting Co., Complaint No. 10073, Reel 214 (3rd District Court, Salt Lake County, 1908)

Charles Steadman vs. U.S. Smelting Co., Complaint No. 10272, Reel 216 (3rd District Court, Salt Lake County, 1908)

Alma Hogenson and May L. Hogenson vs. U.S. Smelting Co., Complaint No. 11601, Reel 235 (3rd District Court, Salt Lake County, 1909)

Walter Steadman vs. U.S. Smelting Co., Complaint No. 12231, Reel 245 (3rd District Court, Salt Lake County, 1910)

U.S. Mining Company vs. John McDonald, Complaint No. 12814, Reel 253 (3rd District Court, Salt Lake County, 1910)

S. A. Parry vs. USSRM Co., Complaint No. 17132, Reel 326 (3rd District Court, Salt Lake County, 1913)

Hans R. Peterson and Sons vs. U.S. Smelting Co., Complaint No. 17167, Reel 327 (3rd District Court, Salt Lake County, 1913)

E. F. Lannberg vs. U.S. Smelting Co., Complaint No. 17736, Reel 336 (3rd District Court, Salt Lake County, 1914)

USSRMCo vs. *Ohio Copper Company, Bingham Central Railway, Metropolitan Trust Company of NY, Bingham Mines Company, NY Trust Co.*, Complaint No. 32218, Reel 550 (3rd District Court, Salt Lake County, 1932)

Joe Mascano vs. U.S. Smelting Co., Complaint No. 48576, Reel 960 (3rd District Court, Salt Lake County, 1931)

Joe Mascano vs. U.S. Smelting Co., Complaint No. 53055, Reel 1137 (3rd District Court, Salt Lake County, 1934)

First Security Trust Company vs. Frederick Jaynes, E. I. Chesney, G. S. Simons and U. S. Fuel Company, Complaint No. 55567, Reel 1223 (3rd District Court, Salt Lake County, 1935)

PUBLISHED SOURCES

Addams, Jane. *Twenty Years at Hull-House* Urbana: University of Illinois Press, 1990.

Agamben, Giorgio. "State of Exception." In *State of Exception*, 1–31. Chicago: University of Chicago Press, 2021.

Alaska Museum. "University of Alaska Museum of the North." Tumblr. *University of Alaska Museum of the North* (blog), January 29, 2015. https://alaskamuseum.tumblr.com/post/109507993040/effie-the-wooly-mammoth -on-a-late-summer-day.

Aldrich, Mark. "Preventing 'The Needless Peril of the Coal Mine': The Bureau of Mines and the Campaign against Coal Mine Explosions, 1910–1940." *Technology and Culture* 36, no. 3 (1995): 483–518.

———. *Safety First: Technology, Labor, and Business in the Building of American Work Safety, 1870–1939*. Vol. 13. Baltimore: Johns Hopkins Press, 1997.

Alexander, Ryan M. "The Spanish Flu and the Sanitary Dictatorship: Mexico's Response to the 1918 Influenza Pandemic." *The Americas* 76, no. 3 (2019): 443–65.

Alexander, T. G. "Irrigating the Mormon Heartland: The Operation of the Irrigation Companies in Wasatch Oasis Communities, 1847–1880." *Agricultural History* 76, no. 2 (2002): 172–87.

Álvarez, R. Huizar. "Carta hidrogeológica de la cuenca del Río de las Avenidas, de Pachuca, Hgo., México." *Investigaciones geográficas* 27 (1993): 95–131.

Amozurrutia, Gonzalo. "El Concepto de indio y el trabajo minero en Latinoamérica en la época colonial: La Rebelión de Real del Monte en 1766." Master's diss., Facultad de Filosofía y Letras, Universidad Nacional Autónoma de México, 2017.

Anaconda Copper Mining Company. "Report of the Anaconda Copper Mining Company." *Report of the Anaconda Copper Mining Company*, 1906–14.

Anderson, Warwick. "Modern Sentinel and Colonial Microcosm: Science, Discipline, and Distress at the Philippine General Hospital." *Philippine Studies* (2009): 153–77.

Arnold, Andrew B. *Fueling the Gilded Age: Railroads, Miners, and Disorder in Pennsylvania Coal Country*. Vol. 2. New York: New York University Press, 2014.

Arrington, Leonard J., and Gary B. Hansen. *The Richest Hole on Earth: A History of the Bingham Copper Mine*. Vol. 11. Logan: Utah State University Press, 1963.

Arrington, L. J. *Great Basin Kingdom: An Economic History of the Latter-day Saints, 1830–1900*. Urbana: University of Illinois Press, 2005.

Arroyo, J. M. M. *Historia y narrativa en el ejido de San Francisco Uruapan (1916–1997)*. Zamora: El Colegio de Michoacán AC, 2002.

Bailey, Lynn Robison. *Old Reliable: A History of Bingham Canyon, Utah*. Tucson, AZ: Westernlore Press, 1988.

Baker, Ellen R. *On Strike and on Film: Mexican American Families and Blacklisted Filmmakers in Cold War America*. Chapel Hill: UNC Press Books, 2007.

Baletti, Brenda, Tamara M. Johnson, and Wendy Wolford. "'Late Mobilization': Transnational Peasant Networks and Grassroots Organizing in Brazil and South Africa." *Journal of Agrarian Change* 8, nos. 2–3 (2008): 290–314.

Ballesteros, V. M. *Síntesis de la Guerra de Independencia en el estado de Hidalgo*. Vol. 1. (Pachuca: UAEH), 2005.

Barrera, Rocio Ruiz. "Breve historia de Hidalgo." Mexico City: Fondo De Cultura Economica, 2000.

Beatty, Edward. *Technology and the Search for Progress in Modern Mexico*. Oakland: University of California Press, 2015.

Bell, Wilson T. *Stalin's Gulag at War: Forced Labour, Mass Death, and Soviet Victory in the Second World War*. Toronto: University of Toronto Press, 2018.

Bernstein, Marvin D. *The Mexican Mining Industry, 1890–1950: A Study of the Interaction of Politics, Economics, and Technology*. New York: State University of New York, 1965.

Besserer, Federico, José Díaz, and Raúl Santana. "Formación y consolidación del sindicalismo minero en Cananea." *Revista Mexicana de Sociología* (1980): 1321–53.

Birn, Anne-Emanuelle. *Marriage of Convenience: Rockefeller International Health and Revolutionary Mexico*. Rochester, NY: University of Rochester Press, 2006.

Brinley, John Ervin. "The Western Federation of Miners." PhD diss., University of Utah, 1972.

Bryan, R. R., and M. H. Kuryla. "Milling and Cyanidation at Pachuca." *Transactions of the American Institute of Mining and Metallurgical Engineers* 112 (n.d.): 722–33.

Buchanan, Thomas C. "Class Sentiments: Putting the Emotion Back in Working-Class History." *Journal of Social History* 48, no. 1 (2014): 72–87.

Bucheli, Marcelo. *Bananas and Business*. New York University Press, 2005.

———. "Multinational Corporations, Totalitarian Regimes and Economic Nationalism: United Fruit Company in Central America, 1899–1975." *Business History* 50, no. 4 (July 2008): 433–54. https://doi.org/10.1080/00076790802106315.

Burke, Edmund. *The Works of the Right Honourable Edmund Burke*. 7th ed. Vol. 9. Boston: Wells and Lilly, 1826.

Burt, Roger, and Sandra Kippen. "Rational Choice and a Lifetime in Metal Mining: Employment Decisions by Nineteenth-Century Cornish Miners." *International Review of Social History* 46, no. 1 (2001): 45–75.

Business Corporations under the Laws of Maine. Making of Modern Law Legal Treatises. Portland, ME: Corporation Trust Company of Maine, 1903.

Cacho Jiménez, Luis Emilio. "La fundación del Sindicato Minero-Metalúrgico." In *Los sindicatos nacionales en el México contemporáneo, Minero-Metalúrgico*, edited by Javier Aguilar, 7–38. México: GV Editores, 1987.

Campbell, Kristine A. "Knots in the Fabric: Richard Pearson Strong and the Bilibid Prison Vaccine Trials, 1905–1906." *Bulletin of the History of Medicine* 68, no. 4 (1994): 600–638.

Caputo, Angela. "East Chicago Residents Fleeing Lead Contamination Find Few Housing Options." *The Chicago Tribune*, October 5, 2016.

Cassigoli, Rossana. *Napoleón Gómez Sada: Liderazgo sindical y cultura minera en México*. México: Miguel Ángel Porrúa, 2004.

Castleman, Barry. "Alice Hamilton and the FBI." *International Journal of Occupational and Environmental Health* 22, no. 2 (April 2016): 173–74. https://doi.org/10.1080/10773525.2016.1156922.

Chandler, Alfred D., Jr. *The Visible Hand*. Cambridge, MA: Harvard University Press, 1993.

Chemin, Eli. "Richard Pearson Strong and the Iatrogenic Plague Disaster in Bilibid Prison, Manila, 1906." *Reviews of Infectious Diseases* 11, no. 6 (1989): 996–1004.

Chomsky, Aviva. *West Indian Workers and the United Fruit Company in Costa Rica, 1870–1940*. Baton Rouge: Louisiana State University Press, 1996.

Christy, Howard A. "Open Hand and Mailed Fist: Mormon-Indian Relations in Utah, 1847–52." *Utah Historical Quarterly* 46, no. 3 (1978): 216–35.

Chung, Hong, Patrick T. Gibbons Lai, and Herbert P. Schoch. "The Management of Information and Managers in Subsidiaries of Multinational Corporations." *British Journal of Management* 17, no. 2 (2006): 153–65.

Church, M. A. "Smoke Farming: Smelting and Agricultural Reform in Utah, 1900–1945." *Utah Historical Quarterly* 72, no. 3 (Summer 2004): 196–218.

Coase, Ronald Harry. "The Nature of the Firm." *Economica* 4, no. 16 (1937): 386–405.

Coase, Ronald Harry. "The Problem of Social Cost." *The Journal of Law and Economics* 3 (1960): 1–44.

Colby, Jason M. *The Business of Empire: United Fruit, Race, and US Expansion in Central America*. Ithaca, NY: Cornell University Press, 2011.

Conell, Carol, and Samuel Cohn. "Learning from Other People's Actions: Environmental Variation and Diffusion in French Coal Mining Strikes, 1890–1935." *American Journal of Sociology* 101, no. 2 (1995): 366–403.

Congress of Industrial Organizations (US), and James B. Carey. *Report of the CIO Delegation to the Soviet Union*. Washington, DC: Congress of Industrial Organizations, 1945.

Connolly, Christopher Powell. *The Devil Learns to Vote, the Story of Montana*. New York: Covici, Friede, 1938.

Connolly, Michael C. *Seated by the Sea: The Maritime History of Portland, Maine, and Its Irish Longshoremen*. Gainesville: University Press of Florida, 2010.

CRB and Mining Outlook. "Copper." February 7, 1917.

Council, National Research. *Industrial Research Laboratories of the United States.* New York: R. R. Bowker Company, 1920.

———. *Industrial Research Laboratories of the United States.* New York: R. R. Bowker Company, 1920.

Cox, Archibald. *Law and the National Labor Policy.* Institute of Industrial Relations, Los Angeles: University of California, 1960.

———. "Revision of the Taft-Hartley Act." *West Virginia Law Review* 55 (1952): 100.

———. "Some Aspects of the Labor Management Relations Act, 1947." *Harvard Law Review* 61, no. 1 (1947): 1–49.

Cox, Archibald, and John T. Dunlop. "Regulation of Collective Bargaining by the National Labor Relations Board." *Harvard Law Review* 63, no. 3 (1950): 389–432.

———. "The Duty to Bargain Collectively during the Term of an Existing Agreement." *Harvard Law Review* 63, no. 7 (1950): 1097–1133.

Crump, Scott. *Bingham Canyon: The Richest Hole on Earth.* Salt Lake City: Great Mountain West Supply, 2005.

Cuenya, Miguel Ángel. "México ante la pandemia de influenza de 1918: encuentros y desencuentros en torno a una política sanitaria." *Astrolabio* 13 (2014): 38–65.

Danks, Noblet Barry. "The Labor Revolt of 1766 in the Mining Community of Real Del Monte." *The Americas* 44, no. 2 (1987): 143–65.

Davies, Brian E., Charlotte Bowman, Theo C. Davies, and Olle Selinus. "Medical Geology: Perspectives and Prospects." In *Essentials of Medical Geology,* edited by Olle Selinus, 1–13. Dordrecht: Springer, 2013.

Davies, Mel. "Cornish Miners and Class Relations in Early Colonial South Australia: The Burra Burra Strikes of 1848–49." *Australian Historical Studies* 26, no. 105 (1995): 568–95.

Davison, Henry Pomeroy. *The American Red Cross in the Great War.* Vol. 46. New York: Macmillan, 1919.

de Grazia, Victoria. *Irresistible Empire: America's Advance through Twentieth-Century Europe.* Cambridge, MA: Harvard University Press, 2009.

Derickson, Alan. "Health Programs of the Hardrock Miners' Unions, 1891–1925." PhD diss., University of California, San Francisco, 1986.

———. *Workers' Health, Workers' Democracy: The Western Miners' Struggle, 1891–1925.* Ithaca, NY: Cornell University Press, 2019. https://doi.org/10.7591/9781501745690.

Dixon, E. James. *Bones, Boats & Bison: Archeology and the First Colonization of Western North America.* Albuquerque: University of New Mexico Press, 1999.

Dombroski, Thomas W. *How America Was Financed: The True Story of Northeastern Pennsylvania's Contribution to the Financial and Economic Greatness of the United States of America.* Bloomington, IN: iUniverse, 2011.

Draper, F. W. "The Story of the Russian Platinum Shipment of 1917." *Mineral Industry* XXVI (1918): 550–55.

Drucker, Peter F. *Concept of the Corporation.* New Brunswick, NJ: Transaction Publishers, 1993.

Dunn, Walter Scott. *The Soviet Economy and the Red Army, 1930–1945.* Westport, CT: Greenwood Publishing Group, 1995.

Dyal, D. H. "Mormon Pursuit of the Agrarian Ideal." *Agricultural History* 63, no. 4 (1989): 19–35.

ESTO. "El Estadio Hidalgo celebra su 25 aniversario." February 14, 2018. https://www .esto.com.mx/335321-noticias-pachuca-aniversario-estadio-hidalgo-hoy-se-cumplen -25-anos-de-la-inauguracion-del-estadio-hidalgo/.

Ewing, Susan. *Resurrecting the Shark: A Scientific Obsession and the Mavericks Who Solved the Mystery of a 270-Million-Year-Old Fossil.* New York: Simon and Schuster, 2017.

Fell, James E., Jr. *Ores to Metals: The Rocky Mountain Smelting Industry.* Boulder: University Press of Colorado, 2009.

Ferleger, L. "Uplifting American Agriculture: Experiment Station Scientists and the Office of Experiment Stations in the Early Years after the Hatch Act." *Agricultural History* 64, no. 2 (1990): 5–23.

Ferrier, Walter Frederick. *Catalogue of a Stratigraphical Collection of Canadian Rocks Prepared for the World's Columbian Exposition, Chicago, 1893.* Vol. 3. Ottawa: Geological Survey of Canada, 1893.

Ford, C. D. "Materialism and Mormonism: The Early Twentieth-Century Philosophy of Dr. John A. Widtsoe." *Journal of Mormon History* 36, no. 3 (2010): 1–26.

Fredona, Robert, and Sophus A. Reinert. "Leviathan and Kraken: States, Corporations, and Political Economy." *History and Theory* 59, no. 2 (2020): 167–87.

Freyn, Henry J. "Impressions and Experiences in the USSR." *The Society of Industrial Engineers Bulletin* (1931): 12–22.

———. "Russian Industry: Can America Afford to Neglect It?" *Class and Industrial Marketing* (January 1931): 42–43, 56–58.

Gabaccia, Donna, Franca Iacovetta, and Fraser Ottanelli. "Laboring across National Borders: Class, Gender, and Militancy in the Proletarian Mass Migrations." *International Labor and Working-Class History* 66 (2004): 57–77.

García, Mario T. *Mexican Americans: Leadership, Ideology, and Identity, 1930–1960.* Vol. 36. New Haven, CT: Yale University Press, 1989.

Glaeser, O. A. "Gold Bullion Mill and Cyanide Plant, Willow Creek, Alaska." *Mining & Scientific Press* 126 (1921): 601–5.

Glaeser, Oscar A. "Compensation for Lung Changes Due to the Inhalation of Silica Dust." *Archives of Industrial Hygiene and Occupational Medicine* 2 (1950): 56.

———. "Ventilation of Small Metal Mines and Prospect Openings." *Transactions of the American Institute of Mining, Metallurgical and Petroleum Engineers* 126 (1937): 221.

Glaeser, Oscar Arthur. "Intermittent Mine Ventilation at the United Verde Mine as an Economy Measure." PhD diss., University of Washington, 1932.

Gohman, Sean M. "It's Not Time to Be Wasted: Identifying, Evaluating, and Appreciating Mine Wastes in Michigan's Copper Country." *IA: The Journal of the Society for Industrial Archeology* (2013): 5–22.

Gomez, Rocio. *Silver Veins, Dusty Lungs: Mining, Water, and Public Health in Zacatecas, 1835–1946*. Lincoln: University of Nebraska Press, 2020.

Gómez Serrano, Jesús. *Aguascalientes: Imperio de Los Guggenheim*. México: FCE, 1982.

Gómez-Vasconcelos, Martha Gabriela. "El volcán Paricutín en el campo volcánico Michoacán-Guanajuato: una revisión." *Ciencia Nicolaita* 74 (2018): 15–30.

Gorostiza, Francisco. *Los ferrocarriles en la Revolución Mexicana*. México: Siglo XXI, 2013.

Gosplan, USSR. "The Soviet Union Looks Ahead: The Five-Year Plan for Economic Construction." New York: H. Liveright, 1929.

Haimson, Leopold H., and Charles Tilly, eds. *Strikes, Wars, and Revolutions in an International Perspective: Strike Waves in the Late Nineteenth and Early Twentieth Centuries*. Cambridge: Cambridge University Press, 2002.

Hamilton, Alice. *Exploring the Dangerous Trades—The Autobiography of Alice Hamilton, MD*. Redditch, UK: Read Books Ltd., 2013.

———. *Industrial Poisons in the U.S.* New York: The Macmillan Co., 1929.

———. *Lead Poisoning in the Smelting and Refining of Lead*. Bulletin of the US Bureau of Labor Statistics. Washington, DC: US Government Printing Office, 1914.

———. "A Mid-American Tragedy." *Sur-Vey Graphic* 29 (1940): 434–37.

———. *Nineteen Years in the Poisonous Trades*. New York: Harper & Bros., 1929.

Hamilton, Alice, and Harriet L. Hardy. "Industrial Toxicology." New York: Harper and Brothers Publishing, 1974.

Hammond, John Hays. "Russia of Yesterday and Tomorrow." *Scribner's Magazine* LXXI (May 1922): 515–27.

Hardt, Michael, and Antonio Negri. *Empire*. Cambridge, MA: Harvard University Press, 2000.

Hawley, Charles Caldwell. *A Kennecott Story: Three Mines, Four Men, and One Hundred Years, 1887–1997*. Salt Lake City: University of Utah Press, 2015.

Hawthorne, Michael. "7 Years Later, New Study Shows East Chicago Kids Exposed to More Lead Because of Flawed Government Report." *The Chicago Tribune*, August 20, 2018.

Heath, Herbert Milton. *Comparative Advantages of the Corporation Laws of All the States and Territories*. Maine: Kennebec Journal Print, 1902.

Heiser, V. G. "Recollections of Cholera at the Turn of the Century." *Bulletin of the New York Academy of Medicine* 47, no. 10 (1971).

Heiser, Victor G. "American Sanitation in the Philippines and Its Influence on the Orient." *Proceedings of the American Philosophical Society* 57, no. 1 (1918): 60–68.

———. "Leprosy in the Philippine Islands." *Public Health Reports* (1896–1970), August 13, 1909.

———. "Reminiscences on Early Tropical Medicine." *Bulletin of the New York Academy of Medicine* 44, no. 6 (1968): 654–60.

Hendrickson, Mark. "The Sesame That Opens the Door of Trade: John Hays Hammond and Foreign Direct Investment in Mining, 1880–1920." *The Journal of the Gilded Age and Progressive Era* 16, no. 3 (2017): 325.

Hijar, Andres. *Where Is Our Revolution?: Workers in Ciudad Juarez and Parral-Santa Barbara during the 1930s.* PhD. diss, Northern Illinois University, 2015.

Hobbes, Thomas. *The Elements of Law, Natural and Politic: Part I, Human Nature, Part II, de Corpore Politico; with Three Lives.* New York: Oxford University Press, 1999.

———. *Leviathan: Or the Matter, Form, and Power of a Commonwealth Ecclesiastical and Civil.* London: G. Routledge and Sons, 1887.

Howe, Mark Antony De Wolfe. *The Harvard Volunteers in Europe: Personal Records of Experience in Military, Ambulance, and Hospital Service.* Cambridge, MA: Harvard University Press, 1916.

Hurley, Andrew. *Environmental Inequalities: Class, Race, and Industrial Pollution in Gary, Indiana, 1945–1980.* Chapel Hill: University of North Carolina Press, 1995.

Hutton, James. *Theory of the Earth: With Proofs and Illustrations.* Vol. 1. Library of Alexandria, 1795.

Indiana, Federal Writers' Project. *The Calumet Region Historical Guide: Containing the Early History of the Region as Well as the Contemporary Scene within the Cities of Gary.* East Chicago: German Printing Company, 1939.

Jeppson, Joseph Horne. *The Secularization of the University of Utah, to 1920.* Berkeley: University of California Press, 1973.

Kennecott Copper Corporation. "Annual Report of the Kennecott Copper Corporation." *Annual Report of the Kennecott Copper Corporation,* 1916–33.

Kime, B. "Exhibiting Theology: James E. Talmage and Mormon Public Relations, 1915–20." *Journal of Mormon History* 40, no. 1 (2014): 208–38.

Kirkpatrick, David. *The Facebook Effect: The Inside Story of the Company That Is Connecting the World.* New York: Simon and Schuster, 2011.

Klubock, Thomas Miller. *Contested Communities.* Durham, NC: Duke University Press, 1998.

Knight, Alan. *The Mexican Revolution: Counter-Revolution and Reconstruction.* Vol. 2. Lincoln: University of Nebraska Press, 1990.

Kobrin, Stephen J. "Expatriate Reduction and Strategic Control in American Multinational Corporations." *Human Resource Management* 27, no. 1 (1988): 63–75.

Kofas, Jon V. "US Foreign Policy and the World Federation of Trade Unions, 1944–1948." *Diplomatic History* 26, no. 1 (2002): 21–60.

Koller, Christian. "Local Strikes as Transnational Events: Migration, Donations, and Organizational Cooperation in the Context of Strikes in Switzerland (1860–1914)." *Labour History Review* 74, no. 3 (2009): 305–18.

Kuryla, Michael H., and Galen Clevenger. "H. Liquid-Oxygen Explosives at Pachuca." *Transactions of American Institute of Mining and Metallurgical Engineers* 69 (1923): 321–40.

Ladd, Doris M. *Génesis y desarrollo de una huelga: Las luchas de los mineros mexicanos de la plata en Real del Monte, 1766–1775.* México: Alianza Editorial, 1992.

Lambert, Josiah Bartlett. *"If the Workers Took a Notion": The Right to Strike and American Political Development.* Ithaca, NY: Cornell University Press, 2005.

Leal, Juan Felipe. *En la revolución (1910–1917).* Vol. 5. México: Siglo XXI, 1988.

Leech, Brian. "Competition, Community, and Entertainment: The Anaconda Company's Promotion of Mine Safety in Butte, Montana, 1915–1942." *The Mining History Journal* 24, no. 1 (2017): 19–39.

Lichtblau, George E. "The World Federation of Trade Unions." *Social Research* 25, no. 1 (1958): 1–36.

Lind, Elizabeth W. "Mills." In *Midvale History, 1851–1979,* edited by Maurice Jensen, 217–23. Midvale, UT: Midvale Historical Society, 1979.

Lindell, J. "Mormons and Native Americans in the Antebellum West." Master's thesis, San Diego State University, Arts and Letters, 2011.

Lindgren, Waldemar. "The Copper Deposits of the Clifton-Morenci District, Arizona." Washington, DC: US Government Printing Office, 1906.

———. "The Gold Belt of the Blue Mountains of Oregon. US Government Printing Office." In *The Gold Belt of the Blue Mountains of Oregon: Twenty-Second Annual Report of the United States Geological Survey to the Secretary of the Interior 1901,* 754–55. Washington, DC: Government Printing Office, 1902.

———. *Mineral Deposits.* New York: McGraw-Hill Book Company, Incorporated, 1913.

———. "The Tin Deposits of Chacaltaya, Bolivia." *Economic Geology* 19, no. 3 (1924): 223–28.

Lindgren, Waldemar, and Clyde P. Ross. "The Iron Deposits of Daiquiri, Cuba." *Transactions of the American Institute of Mining Engineers* 53 (1916): 40–66.

Lindgren, Waldemar, and Edson Sunderland Bastin. "The Geology of the Braden Mine, Rancagua, Chile." *Economic Geology* 17, no. 2 (1922): 75–99.

Lingenfelter, Richard E. *The Hardrock Miners: A History of the Mining Labor Movement in the American West, 1863–1893.* Berkeley: University of California Press, 1981.

Lorence, James J. Palomino. *Clinton Jencks and Mexican-American Unionism in the American Southwest.* Urbana: University of Illinois Press, 2013.

Lucier, Paul. *Scientists and Swindlers: Consulting on Coal and Oil in America, 1820–1890.* Baltimore: Johns Hopkins University Press, 2008.

Lyell, Charles. "On Certain Trains of Erratic Blocks on the Western Borders of Massachusetts, United States," 86–97. London: Proceedings at the Meeting of the Members of the Royal Institution of Great Britain, 1855.

———. "On Foot-Marks Discovered in the Coal-Measures of Pennsylvania." *Quarterly Journal of the Geological Society* 2, nos. 1–2 (1846): 417–20.

———. "On the Relative Age and Position of the So-Called Nummulite Limestone of Alabama." *Quarterly Journal of the Geological Society* 4, nos. 1–2 (1848): 10–17.

———. "On the Structure and Probable Age of the Coal-Field of the James River, near Richmond, Virginia." *Quarterly Journal of the Geological Society* 3, nos. 1–2 (1847): 261–80.

Lynch, Thomas. "The 'Domestic Air' of Wilderness: Henry Thoreau and Joe Polis in the Maine Woods." *Weber Studies* 14, no. 3 (1997): 38–48.

Lynd, Staughton. "Government without Rights: The Labor Law Vision of Archibald Cox." *Industrial Relations Law Journal* 4, no. 3 (1981): 483–95.

MacMillan, Donald. *Smoke Wars: Anaconda Copper, Montana Air Pollution, and the Courts, 1890–1924*. Helena: Montana Historical Society, 2000.

Maitland, Frederic W. "Crown as Corporation." *Law Quarterly Review* 17 (1901): 131–46.

Malone, Michael P. *The Battle for Butte: Mining and Politics on the Northern Frontier, 1864–1906*. Seattle: University of Washington Press, 2012.

Marcosson, Isaac Frederick. *Anaconda*. New York: Dodd, Mead, 1957.

———. *Metal Magic: The Story of the American Smelting & Refining Company*. New York: Farrar, Straus, 1949.

Márquez Morfín, Lourdes, and América Molina del Villar. "El Otoño de 1918: Las Repercusiones de La Pandemia de Gripe En La Ciudad de México." *Desacatos* no. 32 (2010): 121–44.

Martin, James W. *Banana Cowboys: The United Fruit Company and the Culture of Corporate Colonialism*. Albuquerque: University of New Mexico Press, 2018.

Marx, Karl. *Capital*. Translated by Ernest Untermann. Vol. I. Chicago: Kerr, 1912.

Matsusaka, Yoshihisa Tak. *The Making of Japanese Manchuria, 1904–1932*. Leiden: Brill, 2020.

Matusow, Harvey. *False Witness*. New York: Cameron & Kahn, 1955.

Mead, Warren. "Report on Geological Studies in the Pachuca—Real Del Monte Silver District." MITDC, *Warren Mead Papers* MC 625 b2.

Mellinger, Philip J. *Race and Labor in Western Copper: The Fight for Equality, 1896–1918*. Tucson: University of Arizona Press, 1995.

Mendizábal, Miguel O. de. "Los Minerales de Pachuca y Real Del Monte En La Época Colonial: Contribución a La Historia Económica y Social de México." *El Trimestre Económico* 8, no. 30 (1941): 253–309.

Merali, Yasmin. "Individual and Collective Congruence in the Knowledge Management Process." *The Journal of Strategic Information Systems* 9, nos. 2–3 (2000): 213–34.

Mogica, J. H. *Organización campesina y lucha agraria en el Estado de Hidalgo, 1917–1940*. Pachuca: UAEH, 2000.

Moralina, Aaron Rom O. "State, Society, and Sickness: Tuberculosis Control in the American Philippines, 1910–1918." *Philippine Studies* (2009): 179–218.

Morin, G., and G. Calas. "Arsenic in Soils, Mine Tailings, and Former Industrial Sites." *Elements* 2, no. 2 (April 1, 2006): 97–101. https://doi.org/10.2113/gselements.2.2.97.

National Mining Congress, ed. *Report of the Proceedings*. Denver, CO.

National Park Service. "Kennecott Story." *National Park Service, US Department of the Interior.* https://www.nps.gov/wrst/learn/historyculture/upload/Kennecottbulletin.pdf.

NLRB. *Case No. R-2001—Decided September 21, 1940.* Vol. 27. Decisions and Orders of the NLRB. Washington, DC: US Government Printing Office, 1942.

Noble, David F. *America by Design: Science, Technology, and the Rise of Corporate Capitalism.* New York: Oxford University Press, 1979.

Novelo, Victoria. "De huelgas, movilizaciones y otras acciones de los mineros del carbón de Coahuila." *Revista Mexicana de Sociología* (1980): 1358–59.

O'Brien, Thomas F. *The Revolutionary Mission: American Enterprise in Latin America.* Vol. 81. Cambridge: Cambridge University Press, 1900.

———. *The Revolutionary Mission: American Enterprise in Latin America, 1900–1945.* Cambridge: Cambridge University Press, 1999.

Ortega Morel, Javier. "Minería y tecnología: La compañía norteamericana de Real del Monte y Pachuca, 1906 a 1947." PhD diss., Facultad de Filosofía y Letras, Universidad Nacional Autónoma de México, 2010.

Pace, Eric. "Charles E. Wyzanski, 80, Is Dead." *New York Times,* section A, September 5, 1986.

Paracelsus, Jolande Jacobi, and Norbert Guterman. *Paracelsus: Selected Writings.* New York: Pantheon Books, 1951.

Paracelsus, Theophrastus. *Four Treatises of Theophrastus von Hohenheim Called Paracelsus.* Baltimore: John Hopkins University Press, 1996.

Pardo, E. Carrington. "Saneamiento general en relación con las enfermedades de origen hídrico en San Miguel Regla, Hgo." PhD diss., Universidad Nacional Autónoma de México, 1947.

Peirce, George. "Report on the Injury to Vegetation Due to Smelter Smoke in the Salt Lake Valley, Utah in 1914." George James Peirce Papers, 1914–15. MLUU.

Pelis, Kim. *Charles Nicolle, Pasteur's Imperial Missionary: Typhus and Tunisia.* Rochester, NY: University of Rochester Press, 2006.

Perkins, Frances. "What You Really Want Is an Autopsy: Opening Remarks of Frances Perkins to the Tri State Silicosis Conference in Joplin, Missouri, 1940." In *Milestone Documents of American Leaders: Exploring the Primary Sources of Notable Americans,* edited by Paul Finkelman, 1691–92. Dallas: Schlager Group, 2009.

Perrin, Tomas G. "Contribución al estudio histopatológico de la silicosis pulmonar en México: Nota primera algunas consideraciones sobre cien exámenes microscópicos." México: Departamento del Trabajo, 1934.

———. "Contribución al estudio histopatológico de la silicosis pulmonar en México." *Gaceta Médica de México,* November 8, 1933.

Perujo de la Cruz, Rodrigo. "Al grito de ¡revoltura!: Rebelión y cultura política en Real del Monte en 1766." Bachelor's thesis, Facultad de Filosofía y Letras, Universidad Nacional Autónoma de México, 2012.

Pizzolato, Nicola. "Workers and Revolutionaries at the Twilight of Fordism: The Breakdown of Industrial Relations in the Automobile Plants of Detroit and Turin, 1967–1973." *Labor History* 45, no. 4 (2004): 419–43.

Powell, Allan Kent. "A History of Labor Union Activity in the Eastern Utah Coal Fields: 1900–1934." PhD diss., University of Utah, 1976.

Press, Steven. *Rogue Empires*. Cambridge, MA: Harvard University Press, 2017.

Price, Langford Lovell. "'West Barbary;' or Notes on the System of Work and Wages in the Cornish Mines." *Journal of the Royal Statistical Society* 51, no. 3 (1888): 494–566.

Probert, Alan. "En pos de la plata." Pachuca: UAEH, 2011, 446.

Quivik, Fredric L. "Landscapes as Industrial Artifacts: Lessons from Environmental History." *IA: The Journal of the Society for Industrial Archeology* 26, no. 2 (2000): 55–64.

Quivik, Fredric Lincoln. *Smoke and Tailings: An Environmental History of Copper Smelting Technologies in Montana, 1880–1930*. Philadelphia: University of Pennsylvania Press, 1998.

Randall, Robert W. "British Company and Mexican Community: The English at Real Del Monte, 1824–1849." *The Business History Review* (1985): 622–44.

———. *Real Del Monte: A British Silver Mining Venture in Mexico*. Austin: University of Texas Press, 1972.

Rees, John D. "Effects of the Eruption of Parícutin Volcano on Landforms, Vegetation, and Human Occupancy." *Volcanic Activity and Human Ecology* (1979): 249–92.

Reeve, W. P. *Making Space on the Western Frontier: Mormons, Miners, and Southern Paiutes*. Urbana: University of Illinois Press, 2010.

Reygadas, Luis. *Proceso de trabajo y acción obrera: Historia sindical de los mineros de Nueva Rosita, 1929–1979*. México: INAH, 1988.

Rice, V. E. "The Arizona Agricultural Experiment Station: A History to 1917." *Arizona and the West* 20, no. 2 (1978): 123–40.

Richter, F. Ernest. "The Amalgamated Copper Company: A Closed Chapter in Corporation Finance." *The Quarterly Journal of Economics* 30, no. 2 (1916): 387–407.

Ringenberg, Matthew, and Joseph Brain. *The Education of Alice Hamilton: From Fort Wayne to Harvard*. Bloomington: Indiana University Press, 2019.

Rosner, David, and Gerald Markowitz. *Deadly Dust: Silicosis and the Politics of Occupational Disease in Twentieth-Century America*. Princeton, NJ: Princeton University Press, 1994.

Ruiz de la Barrera, Rocío. "La Empresa de Minas Del Real Del Monte (1849–1906)." PhD diss., México: El Colegio de México, 1995.

Ruiz Medrano, C. R. "El tumulto de abril de 1757 en Actopan: Coerción laboral y las formas de movilización y resistencia social de las comunidades indígenas." *Estudios de Historia Novohispana* no. 36 (2007): 108–9.

Ryle, Gilbert. *The Concept of Mind*. New York: Routledge, 2009.

Salinas-Rodríguez, Eleazar, Juan Hernández-Ávila, Isauro Rivera-Landero, Eduardo Cerecedo-Sáenz, María Isabel Reyes-Valderrama, Manuel Correa-Cruz, and Daniel Rubio-Mihi. "Leaching of Silver Contained in Mining Tailings, Using Sodium Thiosulfate: A Kinetic Study." *Hydrometallurgy* 160 (March 1, 2016): 6–11. https://doi.org/10.1016/j.hydromet.2015.12.001.

San Salvador, Gobierno. *Reclamo a Morrill B. Spaulding*. San Salvador: Imprenta Nacional, 1903.

Santiago, Myrna L. *The Ecology of Oil: Environment, Labor, and the Mexican Revolution, 1900–1938*. Cambridge: Cambridge University Press, 2006.

Sariego, Juan Luis. *Enclaves y minerales en el norte de México: Historia social de los mineros de Cananea y Nueva Rosita, 1900–1970*. México: CIESAS, 1988.

Sariego, Luis. *El Estado y la minería mexicana: Política, trabajo y sociedad durante el siglo XX*. SEMIP, 1988.

Schell, William. *Integral Outsiders: The American Colony in Mexico City, 1876–1911*. Lanham, MD: Rowman & Littlefield, 2001.

Schmitt, Carl. *The Nomos of the Earth*. New York: Telos Press, 2003.

———. *Political Theology: Four Chapters on the Concept of Sovereignty*. Chicago: University of Chicago Press, 2005.

Seltzer, Curtis. *Fire in the Hole: Miners and Managers in the American Coal Industry*. Lexington: University Press of Kentucky, 1985.

Semerad, Tony. "Zion's Bank Breaks Ground on Midvale Campus." *Salt Lake Tribune*, August 19, 2020.

Serrano, A. Ochoa. *Los agraristas de Atacheo*. Zamora: El Colegio de Michoacán AC, 1989.

Servos, John W. "The Industrial Relations of Science: Chemical Engineering at MIT, 1900–1939." *Isis* 71, no. 4 (1980): 531–49.

Shattuck, George. "Work in Serbia." In *The Harvard Volunteers in Europe: Personal Records of Experience in Military, Ambulance, and Hospital Service*, edited by Mark Antony De Wolfe Howe, 61–70. Cambridge, MA: Harvard University Press, 1916.

Smith, Joseph. 2 Ne. 5:19–24. In *The Book of Mormon: An Account Written by the Hand of Mormon upon Plates Taken from the Plates of Nephi*. Salt Lake City: The Church of Jesus Christ of Latter-day Saints, 1921.

Solorzano, Armando. "Sowing the Seeds of Neo-Imperialism: The Rockefeller Foundation's Yellow Fever Campaign in Mexico." *International Journal of Health Services* 22, no. 3 (1992): 529–54.

Soluri, John. *Banana Cultures: Agriculture, Consumption, and Environmental Change in Honduras and the United States*. Austin: University of Texas Press, 2005.

Spencer, Arthur Coe, and Charles Will Wright. *The Juneau Gold Belt, Alaska*. Washington, DC: US Government Printing Office, 1906.

Springborg, Patricia. "Leviathan, the Christian Commonwealth Incorporated." *Political Studies* 24, no. 2 (1976): 171–83.

Spude, Robert L. *Science and Technology in Alaska's Past*. Anchorage, AK: Alaska Historical Society, 1990.

Spude, Robert Lester. *To Test by Fire: The Assayer in the American Mining West, 1848–1920*. Urbana: University of Illinois Press, 1989.

"Stop Silicosis." US Department of Labor, 1938.

Striffler, Steve. *In the Shadows of State and Capital*. Durham, NC: Duke University Press, 2001.

Striffler, Steve, Mark Moberg, Gilbert M. Joseph, and Emily S. Rosenberg, eds. *Banana Wars: Power, Production, and History in the Americas*. Durham, NC: Duke University Press, 2003.

Strong, Richard P., George C. Shattuck, Max Theiler, Loring Whitman, Joseph C. Bequaert, Glover M. Allen, David H. Linder, and Harold J. Coolidge. *The African Republic of Liberia and the Belgian Congo: Based on the Observations Made and Material Collected during the Harvard African Expedition 1926–1927*. New York: Greenwood Press Publishers, 1930.

Summers, William C. *The Great Manchurian Plague of 1910–1911: The Geopolitics of an Epidemic Disease*. New Haven, CT: Yale University Press, 2012.

NobelPrize.org. "The Nobel Prize in Physiology or Medicine 1928." Accessed June 16, 2022. https://www.nobelprize.org/prizes/medicine/1928/nicolle/lecture/.

Thoreau, Henry David. *The Maine Woods: A Fully Annotated Edition*. New Haven, CT: Yale University Press, 2009.

Todd, Arthur Cecil. *The Cornish Miner in America: The Contribution to the Mining History of the United States by Emigrant Cornish Miners—the Men Called Cousin Jacks*. Vol. 6. Glendale, AZ: Arthur H. Clark Company, 1967.

Toole, Kenneth Ross. "The Anaconda Copper Mining Company: A Price War and a Copper Corner." *The Pacific Northwest Quarterly* 41, no. 4 (1950): 312–29.

United States Congress, Senate, Committee on Foreign Relations, and Subcommittee on Multinational Corporations. *Multinational Corporations and United States Foreign Policy. Hearings, Ninety-third Congress [Ninety-fourth Congress, Second Session]*. Vol. 4. Washington, DC: US Government Printing Office, 1973.

United States Congress, Senate, Committee on the Judiciary. *Amendment of Sherman Antitrust Law: Hearings before the United States Senate Committee on the Judiciary, Sixtieth Congress, First Session, on Apr. 23, 1908*. Washington, DC: Government Printing Office, 1971.

USF Co. *Thirty Years of Coal*. Salt Lake City, 1946.

USSRM Co. *Annual Report*, New York.

———. *Smelting History*. USHS, MSS A-6028 c.1.

———. *United States Smelting, Refining, and Mining Company . . . : Preferred Stock, Common Stock*. Boston: USSRM, 1916.

Veblen, Thorstein. *The Engineers and the Price System*. Vol. 31. New Jersey: Transaction Publishers, 1921.

Vergara, Angela. *Copper Workers, International Business, and Domestic Politics in Cold War Chile*. University Park: Penn State Press, 2010.

VIVE, for Pachuca. "Estadio Hidalgo cumple 29 años; así fue su construcción e inauguración." February 14, 2022. https://www.pachucavive.com/web/estadio-hidalgo-cumple-29-anos-asi-fue-su-construccion-e-inauguracion/.

Walker, William H. "The Technology Plan." *Science* 51, no. 1319 (1920): 357–59.

Walker, Wm. H. "The Spirit of Alchemy in Modern Industry." *Science* 33, no. 859 (1911): 913–18.

Werner, Abraham Gottlob. *New Theory of the Formation of Veins: With Its Application to the Art of Working Mines*. London: A. Constable, 1809.

Widtsoe, J. A. "Joseph Smith as Scientist: A Contribution to Mormon Philosophy." Salt Lake City: General Board, 1908.

———. *The Relation of Smelter Smoke to Utah Agriculture*. Logan, UT: Experiment Station of the Agricultural College of Utah, 1903.

Widtsoe, J. A., and L. A. Merrill. "Lead Ore in Sugar Beet Pulp." *Bulletin of the Utah Agricultural Experiment Station* no. 74 (1902).

Widtsoe, John Andreas. *Rational Theology: As Taught by the Church of Jesus Christ of Latter-Day Saints*. Salt Lake City: Deseret News, 1915.

Williams, Henry S., and Herbert E. Gregory. "Contributions to the Geology of Maine." *Bulletin* no. 165 (1900).

Winchell, H. V. "Geology of Pachuca and El Oro, Mexico." *Transactions of the American Institute of Mining Engineers* 66 (1921): 27–40.

Wisser, Edward Hollister. "Some Applications of Structural Geology to Mining in the Pachuca-Real Del Monte Area, Pachuca Silver District, Mexico." *Economic Geology* 41, no. 1 (1946): 77–86.

Wolf, Frederick, Bruce Finnie, and Linda Gibson. "Cornish Miners in California: 150 Years of a Unique Sociotechnical System." *Journal of Management History* 14, no. 2 (2008) 144–60.

Wyman, Mark. *Hard Rock Epic: Western Miners and the Industrial Revolution, 1860–1910*. Berkeley: University of California Press, 1989.

Wyzanski, Charles E. "The Open Window and the Open Door: An Inquiry into Freedom of Association." *California Law Review* (1947): 336–51.

Index

Civil Rights Congress, 165

Clark, John, 156–157, 160, 162

Clark, William A., 22, 23

class solidarity, 59

Clement, Victor, 25, 35

Cleveland, Newton, 41

Clevenger, Galen H., 80

Clifton Morenci, Arizona, 76

Cloete, Coahuila, 160

Club House, 119

Coahuila: coal mines in, 38, 95; strikes in, 57, 58, 160; US investments in, 30

coal and coal ores, 99, 133–134

coal miner's nystagmus, 91

coal mines: fossils found in, 127; hiring agents for, 61, 213n39; Mexican, 38; union construction in, 47

Coase, Ronald, 15

Coasian terms, 170

coded script of behavior, 135

coding and encryption of information, 121–222

Coeur d'Alene, 152

Cold War era, 200

collective agreements, 141, 163–164

collective bargaining, 163–164

collective contract #2, 141, 142

collective contract #5, 149

collective contract #9, 155–156, 196

collective contracts, 144, 155

collective corporate consciousness, 199

collective unity, conception of, 125

Colombia, 39

colonial powers, 6–7, 8

colonial projects, 101

colonized people, 89

Colorado, 114, 152

Commerce Code (Código de Comercio), 1889, 20

commercial compounds, 34

commercializing networks, 49

Committee for Peaceful Alternatives, 165

Committee for the United States Company, 150

Committee of Industrial Organizations (CIO): anti-communist sectors, 156; founding of, 143; IUMMSW expelled by, 157, 159; labor agreements, 153; leadership, 149; Mexican and Mexican American organization under, 157; World Federation of Trade Unions organization, role in, 154

Committee on Business Corporations, 103

commonwealth, covenants instituting, 11–12

Commonwealth of Massachusetts, 17

communications behemoths, 200

communism, 158, 164

Communist Domination of Union Officials in Vital Defense Industry, 162

Communist International, 106

Communist Party, 158, 162, 164, 165

communists: leaders, 141, 150, 160; organizations suspected of including, 162; reporting of, 147; swearing not to be, 157; workers, alleged, 165

community construction, 126

community life, 200

Compañía de los Aventureros de las Minas del Real del Monte y Pachuca, 27

Compañía de Real del Monte y Pachuca: Anglo management of, 116–117, 192; associations joined by, 188; Cornish management of, 46; dams operated by, 188, 189; history of, 130; incorporation of, 19; Martínez de Rio, P. involvement in, 28–29; protests against operation of, 180; remunerations in, 83; sale of, 200; as subsidiary, 37; US Company acquisition of, 29–30, 45, 177–178; water rights, 177, 191

Compañía de Transmisión Eléctrica de Potencia, 177, 178, 179, 190, 191

smelters: in Chile, 4; construction of, 24,
25; in Mexico, 28, 35; in North America, 42, 51, 173, 175, 202; parent company subsidiaries, dependence on, 34;
wartime production, 186
Smelters Association, 23
smelter workers, 62, 63, 194
smelting companies, 19
smelting corporations, 20
smelting process, 80, 81
Smith, Jack, 40
Smith, Jenson, 58
Smith, Joseph, 181, 183, 231n41
Smith, Joseph F. (nephew of Joseph
Smith), 181
Smith, W. C., 79
smoke farmers (term), 184
Smoot, Reed, 182
snake elimination, 127
Snake River, 120
socavón (tunnel), 26
social actors, 10, 12
social conflict and coordination, 9
social interactions, corporation coordination of, 7–8
socialist economy, 133
socialist organizations, 48
social media giants, 5
social network behemoths, 200
social result, Leviathan as, 12
Sociedad Anónima (Limited Liability
Corporation), 20
Sociedad Unificadora de Los Pueblos de
la Raza Indígena (Unifying Society of
the Peoples of the Indigenous Race),
187
Society of Refining Companies, 189
Society of the American Colony, 117
Society of the Friends of the Soviet
Union in Mexico, 149
sodium, 169
sodium arsenide, 185

sodium arsenite, 34
sodium cyanide, 190
soil, poisoning of, 186
Solomon River, Alaska, dredge on, 41
Sonora, 30, 108, 150
Sonora Placer Company, 38
South (countries south of U.S.), 26, 126
South (U.S.), 26, 87
South Africa, mines in, 96
South African Association for the
Advancement of Science, 36
South African Association of Engineers,
36
South America, 36
South Chicago, lead and manganese pollution in, 201
Southern Australia, 47
Southern Pacific Railway Company, 108
Southern Utah Railroad, 38
South Manchura Railway Company, 101
sovereign body, 12, 13, 109
sovereign entity, 10
sovereign power, 12, 140
sovereign(s): corporation as body of, 14,
105; defining characteristic of, 170;
firms' capacity to occupy space of old,
202; power of, 12
sovereignties, multiple competing, 8
sovereignty (defined), 170
sovereignty of organization: general
structure of, 10; hidden nature of, 16;
multiple forms of, 15; political deci-
sion/definition regarding, 15; wartime,
147; worker challenge to, 151
Soviet engineering students, 133
"soviet of engineers" (term), 115
"soviet" of North American engineers,
109
Soviet of Technicians (term), 106–107
"soviet of technicians" (term), 14
Soviet Russia, 106
Soviet Union: Africa, influence in, 165;

transformation into, 50; military research on, 82; ores, 14, 78; properties, sale of, 197; separation of, 80, 81; smoke from, 184–185; at tourism sites, 202; toxic, 59; transformation into refined metal, 168–169

sulfur: as industrial poison, 89; medical use of, 85, 87, 88, 89; and metals, interaction between, 77, 167; and metals, separation of, 169; ores, 184; physiological effects of, 94; research on, 77; transfer from mineral to human bodies, 105; volcanic eruptions releasing, 193

sulfur dioxide: chemical processes, 169; as disinfectant, 185–186; expulsion of, 80; liberation of, 190; as pesticide, 85; physiological effect of, 94, 185; reactions with water, 173

sulfuric acid: emissions reduction measures, 174, 175; liberation of, 93–94; production of, 92, 107, 169; volcanic eruptions followed by, 193

sulfurous gases, 192

sulfur oxides, 169, 175, 186

sulfur trioxide, 169

Sunnyside, 120

Sunnyside coal mine, 61, 213n39

Sunnyside Orchestra, 119

supplies, fluxes of, 115

supplies divisions, 198

surface plants, 139

surgeons, miners compared to, 75

Sutton, H. T., 89–90, 91

Swain, Robert, 185, 186–187

switchmen, 56

syndicates: Alaska Syndicate, 40; British, 37; Mexican, 142, 143, 147–148 (*see also* Sindicato Industrial de Trabajadores Mineros, Metalúrgicos y Similares de la República Mexicana (SITMRM)); Willows Copper Argentiferous, 36

System of National Waters, 191

T

Taft, Robert, 141

Taft-Hartley Act, 157, 164

tailings: discharge of, 189; dispersion, prevention of, 180; dust from, 139; extraction of, 190; growing piles, 138; long-term effects of, 201, 202, 234n11; pollution from, 179; processing, 188

Ta-Ku-Shan, 101

Talmage, James, 181, 182, 183

Tamaulipas, 33

Tampico port, 33

Tanana River, 64

tax collection, 5

technical elites, 26, 115, 133

technical frontier, 114–115

technicians: body of self-regulated, 125; circulation of, 115, 198–199; concentrates, shipments handled by, 110; incorporation into firm, 14; migration of, 114; mining behemoth power extended by, 42; power, taking, 131; regime of, 120–121; rise of, 23; sovereign body regulated by, 109; Soviet of, 106–107; strikes impact on, 56

technocratic global bubble, 114–115

technology: circulation of, 198–199; exchange of, 114; fluxes of, 115; industrial and wartime uses of, 102–103

Technology Plan, 82

techno-political body, 122

Tecolote Copper Corporation, 38

Tegucigalpa, Honduras, 39

tellurium, 80

Temascaltepec, Estado de Mexico, 72

Terreros, 54

territorial intervention, 8

Tertiary age, 74

Texas oil, 198

Texas Pacific Land Trust, 197

Théodore, Davilmar, 131

Thoreau, Henry David, 17, 18

Thornberg, C. L., 100

tiles, 89–90

tin, 76, 131

Tintic, Utah, 144

Tiro Alto mine, 69

TNT, 79

Tooele, 144, 186

tourism from US, 200, 202

towns, local histories of, 43

toxic waste, 4

Trade Chambers Confederation, 155

trading firms, 8, 11

trails, exploring and photographing, 128

transnational history, 16

transnational workers' organizations, 49, 210n13

Trans-Siberian Railroad, 107

Travis, Maurice, 157, 160, 162

Treat, Joseph, 19

treaty of 1819, 17

tribute (defined), 46

tribute system, disappearance of, 48, 54

tri-state area (Missouri/Oklahoma/ Kansas): activism in, 160; health problems in, 138–139; lead processing in, 92; in Second World War, 150

Tri-State Committee, 140

Tri-State Conference, Joplin, Missouri, 136, 138, 140

Tropical Health Department (Harvard University), 83–89

tropics, 101, 127

Trujillo Gurría, Francisco, 155

tuberculosis: development of, 73; in Huasca Valley, 201; outbreaks of, 87; research on, 90; silicosis, differentiating from, 98, 138; sufferers, health care for, 140

Tunis, 89

tunnels, 54–55, 138

tut (defined), 46

tut and tribute system of remuneration, 46, 47

tutwork, 54, 55

Twain, Mark, 22

typhoid outbreak, 1902, 90

typhus, 83, 84–85, 90, 218–219n53

U

UAF Alumni Association, 129

underground cities, 50

Underground Leviathan, consciousness of being, 16

underground mine centralization, 31, 190

underground rivers, 168

underground water, 178

unemployed, data on, 69

Unifying Society of the Peoples of the Indigenous Race (Sociedad Unificadora de Los Pueblos de la Raza Indígena), 187

Union de Mecánicos, Electricistas y Similares de Parral, 143

Unión de Mecánicos Mexicanos (UMM), 56, 57, 58, 62, 141, 142

Unión General de Obreros y Campesinos de México (OGOCM), 159–160

union leaders and organizers: arrests of, 144–145, 147; in Mexico, 57; opposition to, 72

unions: affiliations in, 69; Ghanian and Soviet compared, 165; international organizing of, 163; limits on freedom of, 164; local actions of, 140, 148, 150; in Maestranza, 57; national, extending structures of, 156; North American integration of, 159; in Pachuca, 141, 142; political bodies organized by, 105; women's role in, 200; writings against, 72–73

United Fruit Company: alliances with, 88; as colonial power agent, 6;

US Oil Company, 25
US Smelting Company, 19, 173
US Smelting Exploration Company of
 Mexico, SA, 37, 38, 208n60
US Southwest, 158, 198
USSR. *See* Soviet Union
U.S.S.R.M. Exploration Company, 19
US Steel Company, 105, 221n98
US Stores Company, 19, 120
US Supreme Court, 20
Utah: biggest hole on Earth dug in, 43;
 economic relevance, expansion of, 23;
 history of, 128–129; mining, environ-
 mental impact in, 172–175; Mormon
 arrival in, 183; *padrones* system in,
 59–60; railway arrival to, 110, 172; in
 Second World War, 150; statehood,
 181; strikes in, 152, 153; technological
 advances in, 132; as territory, 23, 128;
 US Company, bargaining with in, 148;
 workers' accidents in, 72, 215n60
Utah Academy of Sciences, 128
Utah Agricultural Experimental Station,
 181, 182, 183, 184
Utah Company, 19
Utah Consolidated Company, 23, 25
Utah Consolidated Mines Company, 143
Utah Consolidated Mining and Smelting
 Company, 174
Utah Copper Company (*later* Ken-
 necott): eight-hours regulation, refusal
 to implement, 143; Kennecott absorp-
 tion of, 40; laboratory, 80; manage-
 ment, 61; mills, 62; ores from, 173;
 porphyry copper from, 25; shares, 35; as
 titan, 25; workers, 61–62
Utah Copper Mining Company, 41
Utah Delaware Mining Company, 143
Utah Federation of Labor, 144
Utah Historical Landmarks Association,
 128
Utah-Idaho Sugar Company, 25

Utah mines, snake elimination inside, 127
Utah Railroad, 38
Utah Railway, 111
Utah Railway Company, 19, 120
Utah Society of Engineers, 131
Utah Supreme Court, 143
Utes, 171–172
Utley Wedge, 78
U.V. Industries, 197, 198

V
Valdes, Miguel Aleman, 96, 220n75
Valenzuela, Gilberto, 58
Valparaiso, 112
value-based remunerations, 47
vanadium, 134
Vanadium mine, 157
Van Law, Carlos W., 30, 83–84
Vargas Lugo, Bartolomé, 201
vascular system, 93
Veblen, Thorstein, 14, 106–107, 109
vecindades, 66
Velasco, 191
Velasco treatment plant, 176, 190
Venezuela, 39–40
Venture Corporation, 29
Veracruz, 30, 95
Verk Isotz Estates, 107
verruga peruviana, 86
Vesuvius Fruit Company (fictional com-
 pany), 6
Veta Paricutín, 193
veta Vizcaína, 26, 27, 31
vice director (of international division),
 110
vice director (US Company subsidiary),
 188–189
vice secretary of interior, 58
vice secretary of labor, 155
Villa, Pancho, 30, 188
Villaseñor, Victor Manuel, 148–149
violence, 52

About the Author

ISRAEL G. SOLARES was born and raised in Zacapan, Xochimilco. He received a BA and master's in economics at Facultad de Economía, Universidad Nacional Autónoma de México, and a master's and PhD in History at El Colegio de México. This is his first book and is the result of the research for his PhD and a variety of research stays in North America.